T0137680

ELECTRE AND DECISION SUPPORT

Methods and Applications in Engineering and Infrastructure Investment

ELECTRE AND DECISION SUPPORT

Methods and Applications in Engineering and Infrastructure Investment

by

Martin Rogers

Michael Bruen

Lucien-Yves Maystre

Kluwer Academic Publishers
Boston/Dordrecht/London

Distributors for North, Central and South America:
Kluwer Academic Publishers
101 Philip Drive
Assinippi Park
Norwell, Massachusetts 02061 USA
Telephone (781) 871-6600
Fax (781) 871-6528
E-Mail <kluwer@wkap.com>

Distributors for all other countries:
Kluwer Academic Publishers Group
Distribution Centre
Post Office Box 322
3300 AH Dordrecht, THE NETHERLANDS
Telephone 31 78 6392 392
Fax 31 78 6546 474
E-Mail <orderdept@wkap.nl>

Electronic Services <http://www.wkap.nl>

Library of Congress Cataloging-in-Publication

Rogers, Martin (Martin Gerard)
Electre and decision support : methods and applications in engineering and infrastructure investment / by Martin Rogers, Michael Bruen, Lucien-Yves Maystre.
p. cm.
Includes bibliographical references.
ISBN 0-7923-8647-7
1. Multiple criteria decision making. 2. Decision support systems. 3. Operations research. I. Bruen, Michael. II. Maystre, Lucien-Yves. III. Title.

T57.95.R64 1999
658.4'03--dc21 99-044531

Printed on acid-free paper.

Printed in the United States of America

Table of Contents

Acknowledgements

We would like to acknowledge the assistance of the following people in the production of this text:

Prof. Bernard Roy for so kindly agreeing to write the foreword, and for his general comments on the text.

Dr Jacques Pictet for his help and assistance in the production of the text.

Ms A. Casey and Ms M. Cronin for their contribution to the case study in Chapter 5.

Foreword

Having personal knowledge of the authors' professional experience and skills, it was with great pleasure that I read this book. This book is the first in English which permits a large audience to discover in detail the ELECTRE methods, and, in broader terms, the practice of Multicriteria Decision Aid Methodology. The three authors are civil engineers with wide experience in the practical application of ELECTRE. Following a very concise presentation of the methods, the authors outline the advantages of using them in practice. They highlight this feature by applying different versions of ELECTRE to three case studies.

The book focuses on the areas of engineering and infrastructure investment. It begins with some general comments about the different decision components within the project planning process - the definition of objectives, the identification of alternative courses of action, the establishing of criteria, the evaluation of alternatives and the final recommendation. The authors highlight the ability of Multicriteria Decision Aid to reconcile the economic, technical and environmental dimensions of the project for its planners. They emphasise the complexity of this process, illustrating the importance of identifying the stakeholders within it, as they greatly influence the definition of the decision criteria. A brief case study illustrates these different aspects. Following a comparison of Cost Benefit Analysis and Multicriteria Decision Aid, the introductory chapter sets out the structure of the book, with four subsequent chapters devoted to the methodology of ELECTRE and three outlining case studies involving different versions of ELECTRE.

The chapters concentrating on the ELECTRE methodology first give an overview of the main MCDA methods before presenting the ELECTRE methods in detail. The non-compensatory nature of the methods is emphasised together with their ability to cope with ambiguity and uncertainty. Within these chapters, the reader will find answers to the following questions: In what context should the ELECTRE methods be chosen? Which version of the methods is most appropriate to apply to a given problem? Another chapter deals with a critical and delicate problem within MCDA - how to adequately assess the role played by each criterion in a given decision problem, and how this translates into an appropriate weighting for each one. The last chapter on methodology presents some accessories which, when used with ELECTRE, can greatly enhance its usefulness in practice.

The three case study chapters deal with decision problems in civil and environmental engineering. Each one covers a different civil engineering discipline and each uses a different version of ELECTRE. The first one uses ELECTRE II and involves selecting the most appropriate location for a wastewater treatment plant in Galway, the largest city on the West Coast of Ireland. In the second case, the decision problem uses ELECTRE III to choose the most appropriate solid waste disposal strategy for the Eastern Switzerland region. This case has been described in a previous book in French by Maystre and Bollinger, but is rewritten and updated here. In the final case, selection of the best route for a motorway through the port area of Dublin City is chosen to illustrate the use of ELECTRE IV. In each of the three cases, the reasons for using the selected version of ELECTRE are detailed. Moreover, each one clearly highlights the way multiple criteria analysis together with its accessories can be integrated into the decision process, thereby contributing to the final decision design.

The book emphasises the level of assistance that can be obtained from the ELECTRE methods in a wide array of real life decision contexts. The authors address the complex nature of such problems, including how one deals with incomparabilities within the problem, the multiplicity of actors, the variety of value systems, imperfect information and most importantly the consequences of our actions. They rightly insist that, in decision problems of interest to them, optimisation techniques are not appropriate. Indeed, the term "aid" for them emphasises the virtual impossibility of providing a truly scientific foundation for an optimal solution/decision. The multicriteria approach, together with the techniques derived from it, is presented as a way of achieving a desirable situation, taking into account the available resources.

The authors detail the decision areas where the ELECTRE methods are of particular use in giving this "aid". The way the methods are presented clearly highlights that they are not automatic procedures aimed at forcing the decision. This helps us realise that they are above all a communication tool between the study team and the various actors taking part in the decision process, which result in the provision of a scientifically based solution that can then be submitted to the relevant actors for approval. The authors emphasise that such solutions depend greatly on the quality of the information used. In the ELECTRE methods, this quality is reflected in the various thresholds which form an integral part of the methods, together with the ease with which sensitivity and robustness analyses can be performed. The methods thus help structure an individual's opinions and recommendations, and facilitate discussions which will enable compromise and consensus among actors having conflicting views.

This book is outstanding in many respects. It should permit, for the first time, a broadly based English-speaking audience, both in academia and in the practice, to discover an area of pure and applied research originally undertaken by French-speaking authors. There have been many publications in French in this area, but, until now, those in English have targeted a relatively specialised audience. I am convinced that the simple, clear and concise style of the authors will make this book accessible to very many readers. No important aspect of the subject is neglected, and the concise nature of this book does not hinder its originality. Last but not least, the manner in which the case studies are described allows the authors not only to demonstrate the validity of the approach and procedures presented, but also to help the reader understand how to apply them in an effective manner.

Bernard Roy
Professor at the University Paris-Dauphine
Scientific Advisor at the Transportation Company for the Paris Region (RATP)

1 INTRODUCTION TO MULTICRITERIA DECISION METHODS

1.1 Decisions and the Project Planning Process

The proper planning of a major engineering system requires a set of procedures to be devised which ensures that available resources are allocated as efficiently as possible in its subsequent design and construction. This involves deciding how the available resources, including manpower, physical materials and finance can best be used to achieve the desired objectives of the project developer. Systems analysis can provide such a framework of procedures in which the fundamental issues of design and management can be addressed (de Neufville and Stafford, 1974). Engineering systems analysis provides an orderly process in which all factors relevant to the design and construction of major engineering projects can be considered. Use of the process has the following direct impacts on the coherent and logical development of such a project:

- The process forces the developer to make explicit the objectives of the proposed system, together with how these objectives can be measured. This has the effect of heightening the developers awareness of his overall core objectives
- It provides a framework in which alternative solutions will be readily generated as a means to selecting the most desired one
- Appropriate methodologies for decision making will be proposed within the process for use in choosing between alternatives.
- It will predict the major demands which will be placed on the facility under examination through the interaction of the various technical, environmental and social criteria generated by the process. These demands are not always detected in advance.

The planning of large engineering projects is a rational process. Within it, appropriate future action by the developer is determined by utilising the available scarce resources in such a way that his aims and objectives are maximised. It is a problem solving process which involves closing the gap between the developers objectives and the current situation by means of the developmental project in question, the 'objectives' being, for example, a more coherent transportation infrastructure, better quality rail service, or a more efficient and cleaner water supply system.

The basic procedure can be represented by five fundamental steps. They constitute the foundation of a systematic analysis, and can be listed as follows:

1. Definition of objectives
2. Formulation of criteria / measures of effectiveness
3. Generation of alternatives
4. Evaluation of alternatives
5. Selection of preferred alternative / group of alternatives

DEFINE OBJECTIVES

No planning process should proceed without an explicit statement of the objectives, goals or overall purpose of the proposed undertaking. All analyses have a set of objectives at their basis. Much of the value of the planning process lies in the identification of a clear set of objectives.

The process will generate different classes of objectives that may be potentially conflicting. For example, within the planning of major transport infrastructure, the designer may have to reconcile the maximisation of economic and technical efficiency with the minimisation of social and environmental impact. These objectives will each have their own merits, and must be considered by its own individual set of criteria.

In an engineering context, the determination of broad objectives, such as the relief of traffic congestion in an urban area or changing the method in which domestic waste is disposed of, is seldom within the design engineer's sole remit. Their setting predominantly takes place at what is termed 'systems planning level' where input is mainly political in nature, with the help and advice of technical experts. The objectives serve to define the 'desired situation' that will transpire as a direct result of the construction of the proposed facilities.

ESTABLISH CRITERIA

Defining the planning problem involves identifying the actual gap between the 'desired situation', as defined by the set of objectives derived, and the current situation, and assembling a range of measures designed to minimise or even close it. The ultimate aim of the process is thus to develop a grasp of the relative effectiveness with which these selected alternatives meet the derived set of objectives. Measures of performance, or criteria, must therefore be determined. They are used as 'standards of judging' in the case of the options being examined. Preferably, each criterion should be quantitatively assessed, but, if, as with some social and environmental criteria, they cannot be assessed on any cardinal scale, it should nonetheless be possible to measure them qualitatively, on some graded, comparative scale.

The selection of criteria for the evaluation of alternatives is of crucial importance to the overall process, because it can influence, to a very great extent, the final design. This selection process is also of value because it decides, to a large degree, the final option chosen. What may be seen as most desirable from the perspective of one set of criteria may be seen as much less so using another set. Thus the selection of the preferred design may hinge on the choice of the criteria for evaluation.

IDENTIFY ALTERNATIVE COURSES OF ACTION

Given that the ultimate end point of the process is to identify a preferred solution or group of solutions, it is logical that the decision maker should invest substantial effort in examining a broad range of feasible options. It would not be feasible subject **all** possible options to a thorough analysis. Moreover, because resources for the analysis are never limitless, the decision maker must always be selective in the choice of options to be considered within the process. The decision maker must pay particular attention to identifying those alternatives that are shown to be most productive in achieving objectives, while ensuring that effort spent on the analysis of a given alternative does not exceed its anticipated benefits. This process should result in the drawing up of a set of alternative proposals, each of which would *reasonably* be expected to meet the objectives stated. There is seldom a plan for which reasonable alternatives do not exist.

EVALUATE THE ALTERNATIVES

The relative merit of each option is determined on the basis of its performance on each of the chosen criteria. Each alternative is aligned with its effects, economic costs and benefits, environmental and social impacts and functional effectiveness. This process is usually undertaken using some form of mathematical model. Selecting the appropriate model for the decision problem under consideration is a key step in the evaluation process. In the case of complex engineering projects, where numerous alternatives exist, and where so many variables and limitations need to be considered, it is at this point in the planning process that the application of decision-aid techniques becomes helpful. Ultimately, people make decisions. Computers, methodologies and other tools do not. But techniques and models in decision aid assist engineers / planners in making decisions

SELECTION / RECOMMENDATION

This is the point at which a single plan or short list of approved plans is adopted as most likely to bring about the objectives agreed at the start of the process. This is the real point of decision making, where a judgement is made on the basis of the results of the evaluation carried out in step 4. As expressed above, because the decision is made by people, value judgements must be applied to the objectively derived results from the decision-aid model within the evaluation process. Political considerations may have to be allowed for, together with the distribution of the gains and loses for the preferred alternatives among a range of incident groups affected by the proposed facilities. The act of selection must thus not be seen solely as a technical problem.

1.2 An Introduction to Multicriteria Decision Aid (MCDA)

In order to guide a decision maker in choosing the most appropriate option at the evaluation stage, a set of rules is required to interpret the criterion valuations for each alternative considered. This set of rules, which can be called an evaluation method, is a procedure which enables the pros and cons of alternative projects to be described in a logical framework, so as to assess their various net benefits. They transform the facets of each proposal, as expressed within the agreed measures of performance or criteria, into statements of net social benefits or community welfare.

The evaluation method must provide an insight into the formal relationships between the multiple aspects of alternatives as expressed in their performance on the relative decision criteria. The challenge is to develop an evaluation procedure appropriate for both the decision problem under consideration and the available information. It must be readily understandable to those involved in the decision process. The set of decision rules at the basis of the evaluation process is of vital importance.

If the principle of optimisation is at the basis of the decision rules used, it can be assumed that different objectives, as stated through their relevant measures of performance, can be expressed in a common denominator by means of trade-offs, so that the loss in one objective can be evaluated against the gain in another. This idea of compensatory changes underlies the traditional cost-benefit analysis. The optimising principle is very elegant, providing an unambiguous tool for the evaluation of alternative strategies on the basis of their contribution to community welfare. In the case of cost-benefit analysis, the contribution of each alternative to community is expressed in monetary terms.

In practice, the optimising principle is rather limited, since the specification of a community welfare function presumes that complete information about all possible combinations of actions, about the relative trade-offs between actions, and about all constraints prevailing in the decision making process.

Given the somewhat limiting nature of these constraints on finding solutions to real life problems, the so-called 'compromise' principle should be considered (Van Delft and Nijkamp, 1977). It assumes the existence of a variety of decision criteria, or a variety of decision makers with different priorities. The principle states that any viable solution has to reflect a compromise between various priorities, while the various discrepancies between actual outcomes and aspiration levels are traded off against each other by means of preference weights. The quality of each option can only be judged in relation to multiple priorities, so that a desired alternative is one which performs comparatively well according to these priorities. The compromise principle is particularly relevant for option evaluation / choice problems leading to multicriteria analyses. Given the potential complexity of the planning process for major engineering projects, such multicriteria methodologies can provide a useful resource for decision makers in the completion of their task

Within the context of the 'compromise' principle, Multicriteria decision aid gives project planners some technical tools in order to enable them to solve a decision problem where several often conflicting and opposite points of view must be taken into account within the decision process. With such complex infrastructural planning problems, in many cases no single option exists which is the best in economic, technical and environmental terms. Hence the optimisation techniques available within operations research, referred to above, are not applicable to this problem type. The word 'optimisation' is inappropriate in the context of this type of decision problem. Furthermore, the word 'aid' within the description of the methodology emphasises the virtual impossibility of providing a truly scientific foundation for an optimal solution / decision. Multicriteria Decision Aid (MCDA) provides tools and procedures to help us attain the 'desired situation', as expressed in the set of objectives, in the presence of ambiguity and uncertainty. However refined our models may be, we must recognise that no amount of data will remove the fundamental uncertainties which surround any attempt to peer into the future. Multicriteria methods, which do not yield a single, 'objectively best' solution, but rather yield a kernel of preferred solutions or a general ranking of all options, are the most readily applicable models to problems of option choice within civil engineering where it is virtually impossible to provide a scientific basis for an optimal solution. Solving such a multicriteria problem is therefore not, as Vincke (1992) put it, searching for some kind of 'hidden truth', but rather helping the decision maker to master the complex data involved in a decision problem in such areas and advance towards a solution. This process involves compromise, and depends to a great extent on the personality of the decision maker, and on the circumstances in which the decision aiding process is taking place. However complete the information, the need for personal judgement and experience in making project planning decisions remains. The development of Multicriteria Decision Aid was dictated by real life problems of the type described within this book.

1.3 Participants in the Decision Making Process

In the context of an engineering project, a decision is arrived at when the choice is made as to which alternative or group of alternatives is deemed most suitable to proceed with. This process must include consideration of the 'do-nothing' option. If the decision is to proceed, the way in which the project is to be physically undertaken must also have been resolved. Such complex decisions, often involving the expenditure of vast amounts of money, are rarely taken by one single individual such as a government minister, a technical expert or an administrator. Even if the final legal responsibility does lie with one specific individual, the decision will only be taken after consultation between this designated individual and other interested parties. For example, in Ireland, the final decision regarding whether a major highway project will proceed is the responsibility of the Minister for the Environment. However, his decision is made only after a consultation process with interested parties has been completed, usually by means of a formal public inquiry at which all affected parties are represented.

Such a decision could, in some cases be the ultimate responsibility of a collection of people such as a cabinet of government ministers or an elected or appointed body. Groups seeking to directly influence the decision maker could be professional representative institutions, or local community groups directly affected by the decision. All these 'actors' are, what Banville (1993) calls **stakeholders** in the decision process. They have a pre-eminent interest in the outcome of the process and will intercede to directly influence it. Also, there are **third parties** to the decision such as environmental and economic pressure groups, who are affected only in general terms by the decision, and do not actively participate in making the decision. Their preferences, however, must be considered.

It is usually one of the stakeholders who is being 'aided' within the decision-aid process is identified and designated as the **decision maker**. The diverse backgrounds and differing perspectives of the various stakeholders mean that not all can benefit directly from the decision-aid process. This chosen stakeholder then plays a critical part in the decision-aid process. It is on his / her behalf that the decision-aid process is applied. In some circumstances, however, the designated decision maker may only be a spokesperson for all the stakeholders and / or third parties. Whatever the relative influence of the various actors is, the process requires that a decision maker be identified, even if the objectives specified by the chosen party are those commonly held or assumed to be commonly held by the entire group of stakeholders.

Although the actual process of decision-aid is sometimes carried out by the designated decision maker himself, it is more usual for it to be undertaken by a separate party who is expert in the field of decision theory. This person, called the **facilitator** or the **analyst**, can work alone or as leader of a team.

The function of the analyst is to explain the mechanics of the chosen model to the decision maker, obtain all required input information, and interpret the results of the model in an easily understandable way. The technical complexity and sophistication of modern decision-aid models underlines the importance of the analyst in ensuring the success of the overall process.

Decision-aiding can thus be seen in terms of the operation of a designated person, called the analyst, who, through his expertise in the use of explicit but not necessarily scientifically rigorous models, assists the decision maker in answering questions which clarify the decision to be made, and aids the decision maker in the recommendation of a coherent and consistent solution.

1.4 Defining Criteria

Decision Aid involves the provision of solutions to questions put forward by actors within a decision process using a specific model. To achieve this, the decision maker often has to establish comparisons on the basis of the evaluation of the alternatives according to several criteria. Each criterion represents the decision maker's preferences according to some point of view. The success of decision aid *crucially* depends upon the manner in which the family of criteria has been constructed.

A criterion can be defined as a tool that allows the comparison of alternatives according to a particular 'point of view'. Most real life decision problems, not only in engineering but in many other fields, involve several criteria. In a multiple criteria approach, the decision maker builds several criteria using several points of view. These points of view represent the different axes along which the various actors or participants in the decision process justify, transform and dispute their preferences. The comparisons deduced from each of these criteria should be interpreted as preferences restricted to the aspects taken into account in the point of view underlying the definition of the criterion. A criterion is thus a type of model, allowing the establishment of preference relationships between alternatives. The care and precision with which this model is constructed is thus crucial to the quality of decision aid.

Certain guidelines should be followed in constructing a criterion in order to ensure that the results of the decision aid process is adhered to by all participants:

- All participants in the decision process should understand and accept the points of view underlying the definition of the various criteria, even if they disagree on their relative importance.
- The method of evaluation on each criterion for each alternative should also be understood and accepted by all actors in the decision process. This involves the analyst searching for a decision model which will be acceptable to all participants.
- The construction of the criterion should take into account the quality of the data associated with it.

Building a set of criteria

Multicriteria decision aid begins with the generation of criteria that should provide a means of evaluating the extent to which each alternative achieves the goal of the 'desired situation', as defined by the set of objectives derived in the first phase of the project planning process (Bouyssou, 1990). Building a set of criteria implies that one has chosen and explicitly described what the 'goal' or 'desired situation' is, and that this determination is understood and agreed by all actors involved in the decision process. A panel of experts and / or a literature search can then be employed to identify the criteria relevant to the problem area under scrutiny.

It is vitally necessary that the criteria chosen represent the desired mission or ultimate goal of the decision makers. One way of ensuring this is to derive the criteria hierarchically from a 'super goal', a unique point of view which encapsulates the properties of the 'desired situation'. Belton and Vickers (1990) advocates such a system, where criteria are constructed through the decomposition of a unique point of view into sub-points of view that are again decomposed relevant points of view are reached. Using the hierarchical approach, one often refers to 'criteria', 'sub-criteria' and 'sub-sub-criteria' depending on the level of the hierarchy. Within Multicriteria Decision Aid, use of the word criteria is restricted to those sub-points at the upper levels of the hierarchy. Within engineering decision problems, those attributes at the lower levels of the hierarchy are usually expressed as parameters and indicators.

It is important that the resulting criteria to be used in the decision aid problem have the following basic properties:

- *Be complete and exhaustive:* All important performance attributes deemed relevant to the final solution must be represented by criteria on the list.
- *Be Mutually Exclusive:* This permits the decision maker to view the criteria as independent entities among which appropriate 'trade-offs' may subsequently be made. This property also helps prevent 'double-counting' through the mutual exclusivity of the criteria.

- *Be restricted to performance attributes of real importance to the decision problem.* This provides a sound starting point for the problem, as the less important / irrelevant / unnecessary criteria can be screened out of the process at the earliest possible stage. (Yoon and Hwang, 1995)

Often when planning projects within engineering, the criteria used as a basis of option choice are built up on the basis of a predefined list of possible criteria contained in a set of published guidelines for the planning of such projects. This process of running through an exhaustive list of possibly relevant criteria, including some while excluding others, can aid the 'scoping' process, resulting in a set of exhaustive, fully relevant and mutually exclusive criteria being available to the decision maker. Past experience on the part of the decision maker in drawing up lists of relevant criteria for similar problem types can also assist this process.

Different Criteria Types

The set of criteria arrived at for a civil engineering problem will inevitably come from various sources and fall into different categories. If the decision problem involved only economic criteria, then each would be measurable in monetary units, and a purely economic analysis would be undertaken using engineering economy models to measure the relative 'worth' of the projects under consideration. Within such an assessment, if non-economic criteria were to be evaluated, their evaluation would be seen to an extent as subsidiary, given that they could not be estimated on the same monetary-based scale.

Within a multicriteria decision problem, however, all criterion types are approached on an equal basis, whether they can be estimated in monetary terms, in quantitative but non-monetary terms or in purely qualitative terms. In general, within a civil engineering planning problem, the criteria introduced fall into three broad categories:

- *economic criteria*: Those attributes which express the economic factors associated with a given project option, most importantly the initial construction cost and subsequent running / maintenance costs.
- *technical criteria*: Expressing the technical / engineering factors which influence the desirability of the option in question, and
- *environmental / social criteria*; Those characteristics which measure the social and environmental desirability / undesirability of the proposed project option.

These different criterion types result in attributes that are measured in different ways, with differing degrees of precision. Those in the first category are all measurable in monetary terms. Criteria from the second category are usually measurable in quantitative though not necessarily monetary terms. Those in the third category are most likely to contain a high proportion of attributes that are, to a greater or lesser extent, intangible. Methodologies within MCDA have the ability to accommodate all the different criterion types within the decision problem on an equal basis.

Whatever process is used to select and assess each individual attribute, the final family of criteria chosen should represent the different aspects of the problem at hand, while avoiding redundancies.

1.5 A Brief Example

As a short illustration of the planning process for an engineering / infrastructure investment project, the assessment of different public transport systems for a large urban centre can be outlined as follows:

The objectives of the proposed system might be

1. to improve the quality of life of the city's inhabitants
2. to promote the proper physical development of the city
3. to maximise the efficiency in the implementation of a given strategy
4. to promote employment in the urban area

The criteria / measures of effectiveness derived from these objectives might include

1. reliability / level of comfort of the given alternative
2. resulting reduction in environmental effects
3. level of improved access to cultural facilities
4. extent of impact on suburban development
5. level of disruption during construction
6. construction and maintenance costs
7. extent of resulting job creation
8. level of improvement in access to all job opportunities
9. extent of consolidation of existing business and industry

Options for consideration might include

1. Introduction of an extensive new high quality bus network
2. Construction of a new light rail transit system
3. Major extension of existing heavy rail network
4. Construction of light rail system plus minor extension to heavy rail system at strategic locations
5. the introduction of a co-ordinated cycle network, along with major upgrading of entire existing public transport network

6. minor renovations to the existing public transport network (do-nothing / do-minimum option)

The effects of each option might then be evaluated on each of the stated criteria using, for instance, a multicriteria model which would require, for each of the given criteria, that each option be scored on a predetermined scale, and that relative importance scores for the set of criteria be agreed.

The selection of the chosen alternative involves the decision maker, who may be the director / chief planner of the relevant urban transportation authority, opting for a course of action on the basis of the preceding evaluation, taking into account the distributional effects of the chosen project on the different communities directly affected by the introduction of the proposed facility. This may involve varying the importance weightings of the criteria to reflect their varying effect on the incidence groups. A sensitivity analysis of the effects of changes in criterion scores for each option on the final choice may also help resolve any selection difficulties.

1.6 Multicriteria Decision Methods in Engineering Planning

Within the context of both publicly and privately funded infrastructure projects, engineers form one of the principal spending professions in the sense that they carry out much of the responsibility for the wise use of resources which provide the material basis of an expanding economy. Decision analysis as applied to the planning of engineering projects is concerned with assessing the overall effect of using resources in order to establish priorities between competing proposals. Its purpose for the engineer is to provide a means of judging the economic, technical and social / environmental merits of alternative schemes, thereby ensuring that available scarce resources shall be expended on that project / group of projects which best achieves these set of agreed goals.

Formerly, such evaluations had their basis solely in traditional economic theory. Within this methodology, evaluating the effects of a given project option involves making a distinction between the purely technical and physical effects of the option under examination and the economic evaluation of these effects (Nijkamp, 1975). For example, in the case of a highway project, the following technical-physical effects might occur:

- changes in land use in the vicinity of the project
- increase in accessibility to the region
- decrease in accident rates
- increase in noise and air pollution levels

Once the technical-physical effects of a given alternative have been assessed, attention must be paid to their economic evaluation. This process is necessary to select the optimum plan from the series of options available. Traditional economic theory assumed the existence of a price system which enables technical-physical effects such as those outlined above to be transformed into monetary values. Such a price mechanism should reflect the net social costs or benefits of the effects of each option considered. In reality, however, direct pricing of the constituents of the effects vector is not possible due to the following:

- the absence of a market system for public investments
- various externalities
- the existence of intangibles, and
- imperfect information regarding the effects of all project options

Therefore, adapted evaluation methods are required. There are **two classes** of adapted evaluation methods:

- Cost Benefit Analysis and its variants
- Multi-criteria Methods

Cost Benefit Analysis

The Cost Benefit Analysis (CBA) methodology is of great significance to this area of study because it constituted the first formal decision model utilised in the planning of major engineering projects. It involves the computation, in an indirect way, of monetary values for the different project outcomes.

The practice of Cost Benefit Analysis can be said to date from the introduction of the Flood Control Act (1936) in the United States. The Act required that the Federal Government improve navigable waters for flood control purposes *if the benefits to whosoever are in excess of the estimated costs*. Although no specific definition of costs or benefits were included, or how the actual calculation was to be made, this piece of federal legislation constituted the first expression, in a context of the evaluation of a developmental project, of the notion of CBA. During the 1950's, the US Federal Government agreed a set of rules for comparing costs and benefits, and noted the need to measure them in a common monetary unit. Projects could then be ranked in terms of their ratio of cost to benefit as a guide to judgement.

The 1960's and 1970's witnessed a rapid expansion in the use of Cost Benefit Analysis as a tool for assessing major engineering / infrastructural projects. These studies included the analysis of the costs and benefits of constructing an underground railway in London - The Victoria Line - by Beesley and Foster (1965), the cost benefit analysis for the London Birmingham Motorway by Coburn, Beesley and Reynolds (1960) and the economic analysis for the siting of the proposed third London airport by Flowerdew (1972). This growth was partly the result of the increased Government involvement in the economy during the post-war period, and partly the result of the increased size and complexity of investment decisions in a modern industrial state. The techniques was specifically developed to bring to account, in explicit terms, the cost to society as a whole rather than the narrower, purely financial costs incurred as a result of the project.

While Cost Benefit Analysis has been found to be a powerful instrument of project evaluation, the basic problem inherent in the use of this methodology by practitioners is the fact that the evaluation of a project must relate to an unambiguous monetary uni-dimensional criterion, since a comprehensive cost-benefit approach requires a transformation of all project option effects into one single monetary dimension. This severe restriction is responsible for many difficulties in the practical application of CBA, as attributes / criteria that cannot be readily transformed into monetary units are omitted from its framework. Given these limitations in the application of CBA, several adaptations have been developed, two prominent ones being the Planned Balance Sheet (PBS) and the Goal Achievement Matrix (GAM). Their underlying methodology is discussed in Chapter 2. Their aim was to include the non-monetary based technical, social and environmental effects of a development project alongside the monetary based ones. A brief description of each, together with their general application to the option choice problems within the physical planning process is discussed briefly below.

Variants of Cost Benefit Analysis

The Planned Balance Sheet was devised by Lichfield (1971) as a means of introducing the rigour of Cost Benefit Analysis into project appraisal within regional and urban planning. He believed that physical planning lacked the rigorous tools for decision making, and that cost benefit techniques could be adapted for this purpose. He defined most urban / regional planning decision problems as multi-sectoral, with many divergent groups affected in many ways which could not always be measured in monetary terms. Whereas, in a decision context, Cost Benefit Analysis under-represented these intangibles by simply referring to them in an accompanying narrative to the main analysis, PBS used the framework of a balance sheet to incorporate this information in the same visual framework as the monetary based effects. Lichfield (1975) applied this methodology to various urban planning development proposals in the San Francisco area within the United States, and also to the comparison of two alternative city plans for Cambridge in the United Kingdom.

Hill's Goal Achievement Matrix (1973) evaluates alternative plans within a matrix format on the basis of how well each achieves a set of predetermined objectives. Because of its evolution from Cost Benefit Analysis, the estimation of costs and benefits are central to the method. However, both are defined in terms of goal achievement. Hill had a particular interest in applying the Goal Achievement matrix methodology to the evaluation of transportation plans, the most prominent case study being the evaluation of alternative transportation plans for Cambridge, England.

While both Hill and Lichfield saw their methods as being more inclusive than CBA, and therefore more consistent with the rational planning model, both nonetheless retain the narrowly based structure at the basis of Cost Benefit Analysis, emphasising, by its very nature, those attributes that can be expressed in monetary terms.

Multicriteria Models

The second class of adapted evaluation techniques starts from a different perspective. Instead of a monetary transformation of all different project option outcomes, the non-monetary evaluation methods attempt to take into consideration the multiple dimensions of a decision problem in a balanced manner. When project option effects, or criterion valuations as they are called, are treated in their own dimensions, the problem arises of how to 'weigh' them against each other. Such weighting procedures depend on the relative priorities or importance levels attached to the various decision criteria for the project alternatives being considered. Such methods are called Multicriteria methods.

The major advantage of a multi-criteria analysis is its capacity to take account of an entire range of differing yet relevant criteria, even if these criteria cannot be related to monetary outcomes. On the basis of this concept of a multidimensional compromise, a series of alternative multi-criteria decision methods have been developed.

Multicriteria Methods therefore approach the decision process from a different perspective, assuming, as a matter of course that criteria arising from different perspectives or points of view will, by their nature be estimated on different measurement scales and according to different methodologies. These methods, by their very nature, place all criteria, monetary or non-monetary, quantitative or qualitative, economic or environmental, on an even footing in the context of the decision problem. The multi-criteria methods most relevant for application to engineering / infrastructure investment problems are referred to within the context of this book - Checklist Methods, Multi Attribute Utility Theory (MAUT), Analytic Hierarchy Process and Concordance / ELECTRE Techniques. A brief description of these models, together with relevant applications is provided immediately below. These models are not put forward within the text as an exhaustive list of currently available models within MCDA. Rather, they serve merely to provide a broad indication of the different MCDA models that have been applied to engineering / infrastructure option choice problems in recent years.

Multi-criteria models all operate on the basis of making a preference decision, be it prioritisation, selection or general evaluation, over the available alternatives, each of which are characterised by multiple, usually conflicting, attributes. Checklists and MAUT are both Index or Trade-off Methods, where individual attributes are transformed into units on a common notional scale, rating them in terms of their common importance, and then manipulating them mathematically in order to compute indices which allow the relative evaluation of alternatives. Checklists have been used by Dee (1973) to assess water resource projects, while Utility Theory, devised by Keeney and Raiffa (1976), has been used to evaluate alternatives in a Land Reclamation and Management Problem in Arizona, U.S., to select a strategy for developing airport facilities in Mexico City, and for choosing a framework for river basin development of the Maumee River in Ohio, Indiana and Michigan, U.S. (Goicoechea, Hansen and Duckstein (1982). The Analytic Hierarchy Process (AHP), devised by Saaty (1977), determines the relative merit of each project alternative from a pairwise analysis of the preference ratings for all combinations of project alternatives, for each criterion / attribute involved. The relative importance of each criterion is also determined from a similar pairwise analysis. The overall result is a ranking of all alternatives on an interval scale. In an engineering project planning context, the Analytic Hierarchy Process was used by Saaty to develop an inter-urban transport network between Sudan's major export outlet, Port Sudan, and the other major urban centres within the country (Saaty, 1988), the decision problem being the choice of projects to best implement the economic development strategy for Sudan. The method has also bee used in the problem of optimum choice of a Coal Using Energy System (CUESS) technology (Saaty, 1988). Concordance techniques use various mathematical functions to indicate the degree of dominance of one project alternative or group of alternatives over the remaining ones. They facilitate comparisons between alternative schemes by ascribing initial weights to decision criteria, and then varying these weights as part of a sensitivity analysis, if their exact value is not known. As with AHP, comparison between alternatives proceeds on a pairwise basis with respect to each criterion, with the analysis based on the degree to which project outcomes and preference weights confirm or contradict the pairwise dominance relationships between alternative proposals. The method examines the degree to which the preference weights are in agreement with pairwise dominance relationships, and the degree at which weighted project outcomes differ from each other. These stages are based on a 'concordance and discordance' set (Nijkamp, 1975). The result is a ranking of all options. The concept of Concordance was formulated by Bernard Roy (1968), whose methodology ELECTRE remains dominant in the field (Roy and McCord, 1996). In an engineering context, the method has been used to analyse the impacts of energy alternatives (Siskos and Hubert, 1988), to help select the best location for an incinerator (Maystre, Pictet and Simos, 1994), and in the site selection for a wastewater treatment plant (Rogers and Bruen, 1997). In the context of infrastructure investment, the method has been utilised to rank options for a major urban motorway project (Rogers and Bruen, 1996), to evaluate alternative suburban line extension projects on the Paris Metro System (Roy and Hugonnard, 1982), and on where best to locate industrial production units (Guigou, 1971). Thus, over the past 25 to 30 years, multicriteria decision methods have been widely applied within the field of major engineering related projects.

Conclusion

Project evaluation and decision making require the precise specification of a set of alternative proposals, together with all relevant decision criteria. Particularly in the context of the planning of engineering and infrastructure investment proposals, there are many relevant objectives which cannot be simply traded off against each other by means of monetary units, as is required by the first adapted evaluation technique. Thus, the evaluation of alternatives on the basis of a single, uni-dimensional decision criterion is likely to be fraught with difficulties (Nijkamp and Van Delft, 1977). Public policy issues relating to the planning of major engineering projects is generally formulated on the basis of numerous economic, political, social, technological and environmental objectives. Explicit trade-offs between often complimentary and conflicting objectives are not always apparent, while the selection process is further complicated by the issue of uncertainty.

Planning for multiple objectives thus requires adjusted theories and evaluation methods. Adapted methods of the first type mentioned above - the traditional monetary evaluation methods - are somewhat limited in this respect. When a decision problem with multiple objectives has been identified, the use of the second category of adapted techniques - multi-criteria evaluation methods - provide a logical and flexible approach, which adheres to the compromise principle referred to above.

The underlying theory of each of the above Multicriteria models is outlined concisely in Chapter 2, together with the reasons why, in the opinion of the authors, the ELECTRE Model is most readily applicable to option choice problems for proposed engineering / infrastructure based projects.

1.7 Chapter Review / Outline of Contents

Within Chapter 1, the centrality of the decision process to the planning of engineering and infrastructure projects is outlined, together with the major participants in the process. Given that the problem involves multiple, often conflicting, criteria, these building blocks are defined in detail, and the necessity for their proper construction is emphasised. The different types of criteria to be dealt with within a developmental project are detailed. Finally, a brief overview of the evolution in the use of decision models for major engineering infrastructural projects is given, along with a brief description of each model identified.

Chapter 2 contains brief outlines of the theoretical background to four different multicriteria decision model types:

- Goal Achievement Matrix
- Index Methods (including checklists and MAUT)
- Analytic Hierarchy Process, and
- The ELECTRE Approach

The chapter concludes by explaining why, in the opinion of the authors, the ELECTRE Approach is the most readily applicable decision model to apply to option choice problems at the planning stage of engineering / infrastructure projects.

The ELECTRE Approach involves not just one model but a suite of models. The application of a given model over one of the others depends on the type of answer required and the quality of data available. Chapter 3 gives the detailed theoretical background to each of the 6 major models with the ELECTRE suite.

Chapter 4 explains the process by which the criterion importance weightings to be used within ELECTRE are derived.

Chapters 5, 6 and 7 detail the application of different versions of ELECTRE to case studies in different fields. Chapter 5 illustrates the application of ELECTRE II to the choice of location of a wastewater treatment plant. Chapter 6 contains an application of ELECTRE III to the selection of a solid waste management strategy. Chapter 7 outlines the application of ELECTRE IV to the route selection problem for a major urban motorway. Finally, Chapter 8 presents some new ideas and techniques of particular use and relevance within the area of multicriteria decision aid.

1.8 References

Bouyssou, D. (1990) 'Building Criteria: A Prerequisite for MCDA'. In *Readings in Multiple Criteria Decision Aid*, pp58-80, ed. Bana e Costa, C., Springer-Verlag

Banville, C., Landry, M.,Martel, J.M., Boulaire, C. (1993) 'A Stakeholder's Approach to MCDA', University Laval, CRAEDO, *Working paper* 93-77.

Beesley, M.E. and Foster, C.D. (1965) 'Victoria Line: Social Benefits and Finances', *Journal of the Royal Statistical Society*, Series A, pp67-88.

Belton, V. and Vickers, S. (1990) 'Use of a Simple Multi-Attribute Value Function Incorporating Visual Interactive Sensitivity Analysis for Multiple Criteria Decision Making'. In *Readings in Multiple Criteria Decision Aid*, pp319-334, ed. Bana e Costa, C., Springer-Verlag

Coburn, T.M., Beesley, M.E., Reynolds, D.J. (1960) *The London - Birmingham Motorway: Traffic and Economics*, Road Research Laboratory, Technical Paper No. 46

de Neufville, R. and Stafford, J. (1974) *Systems Analysis for Engineers and Managers*, McGraw Hill

Dee, J. (1973) 'Environmental Evaluation System for Water Resources Planning' *Water Resources Research*, Volume 9, pp523-535.

Flowerdew, A.D.J., (1972) 'Choosing a Site for the Third London Airport: The Roskill Commission Approach' In *Cost Benefit Analysis*, R. Layard (ed), Penguin.

Goicoechea, A., Hansen, D.R., Duckstein, L. (1982) *Multiobjective Decision Analysis with Engineering and Business Applications*, John Wiley and Sons.

Guigou, J.L. (1971) 'On French Location Models for Production Units'. *Regional and Urban Economics,* Volume 1, No. 2, pp107-138.

Hill, M. (1973) *Planning for Multiple Objectives: An Approach to the Evaluation of Transportation Plans*. Technion.

Keeney, R. Raiffa, H. (1976) *Decisions with Multiple Objectives: Preferences and Value Trade-Offs*. John Wiley and Sons.

Lichfield, N. (1971) 'Cost Benefit Analysis in Planning: A Critique of the Roskill Commission' In *Regional Studies*, Volume 5, pp157-183.

Lichfield, N., Kettle, P., Whitbread, M. (1975) *Evaluation in the Planning Process*. Oxford: Pergamon.

Maystre, L., Pictet, J., and Simos, J. (1994) *Methodes Multicriteres ELECTRE*. Presses Polytechniques et Universitaires Romandes, Lausanne.

Nijkamp, P. (1975) 'A Multicriteria Analysis for project Evaluation: Economic - Ecological Evaluation of a Land Reclamation Project'. *Papers of the Regional Science Association*, Volume 35, pp87-111.

Rogers, M.G. and Bruen, M.P. (1996) 'Using ELECTRE to Rank Options within an Environmental Appraisal - Two Case Studies'. *Civil Engineering Systems*, Vol. 13, pp203-221.

Rogers, M.G. and Bruen, M.P. (1997) 'Applying ELECTRE To an Option Choice Problem Within an Environmental Appraisal - Three Case Studies from the Republic of Ireland'. *Proceedings of the International Conference on Methods and Applications of Multicriteria Decision Making*, May 14th to 16th, Mons, Belgium.

Roy, B. (1968) 'Classement et choix en presence de points de vue multiples (la methode ELECTRE)'. *Revue Francaise d'Automatique Information et Recherche Operationelle (RIRO)*. Volume 8, pp57-75.

Roy, B. and Hugonnard, B. (1982) 'Ranking of Suburban Line Extension Projects on the Paris Metro System by a Multi-criteria Method' *Transportation Research Record*, Volume 16A, No. 4, pp301-312.

Roy, B. and McCord, M.R. (1996) *Multicriteria Methodology for Decison Aiding,* Kluwer Academic Publishers.

Saaty, T. (1977) 'A Scaling for Priorities in Hierarchical Structures'. *Journal of Mathematical Psychology*, Volume 15, pp234-281.

Saaty, T. (1988) *The Analytic Hierarchy Process: Planning, Priority Setting, Research Allocation*, McGraw Hill.

Siskos, J., Hubert, P. (1988) 'Multi-criteria Analysis of the Impacts of Energy Alternatives: A Survey and a New Approach'. *European Journal of Operational Research*, Volume 13, pp278-299.

Schwarz, R.M. (1994) *The Skilled Facilitator*, Jossey-Bass Publishers, San Francisco.

Van Delft, A., Nijkamp, P. (1977) *Multi-criteria Analysis and Regional Decision-Making*. Martinus Nijhoff Social Services Division, Leiden.

Vincke, P. (1992)*Multi-criteria Decision Aid*. John Wiley, Chichester, United Kingdom.

Yoon, K.P., Hwang, C-L (1995) Multiple Attribute Decision Making: An Introduction, Sage Publications.

2 MULTI-CRITERIA EVALUATION METHODS

2.1 Introduction

There are a large number of multi-criteria methodologies for choosing between options for engineering and infrastructure investment projects. The main ones are:

- Checklist Methods,
- Multi-Attribute Utility methods (MAUT),
- Analytic Hierarchy Process (AHP),
- Concordance Analysis

The theoretical basis for each of these methods is outlined in this chapter and their advantages and disadvantages explained.

2.2 Checklist Methods

General

Both Checklists and Multi-Attribute Utility Theory (MAUT) can be classified as Index Methods, which transform the individual criteria, each measured in their own set of units, into numbers on a common notional scale, weighting them in terms of their relative importance and deriving an overall score for each option. Index methods can make the decision maker's task relatively straightforward and can be particularly useful when the decision involves a complex variety of evaluation criteria to be assessed for each option.

Basic Methodology

Decision-focused Checklists are widely applied to the assessment of engineering and infrastructure projects, particularly where a high proportion of the criteria are environmentally related (Canter, 1996). Checklists have their theoretical basis in trade-off analysis, which typically involves the comparison, in matrix form, of a set of project options relative to a set of decision factors. The information is displayed in matrix form, with each specific option analysed on the basis of a set of economic, technical, social and environmental criteria. A number of approaches can be utilised to complete the trade-off matrix:

Quantitative / Qualitative Approach
Qualitative and / or quantitative information is provided in a synthesised and integrated form for each project option on each of the decision criteria, and is presented in matrix form. No numerical procedures are carried out on the information. The approach is presentational in nature. This model is called a simple checklist.

Ranking, Rating or Scaling Approach
Here, the decision information for each project option, be it in qualitative or quantitative form, is assessed through assigning it a rank, rating or scale value for each relative decision criterion. The complete set of values is presented within the matrix. This model is called a scaled checklist.

Weighting Approach
Only the relative importance weight of each criterion is considered, and the information on each project option, be it qualitative, quantitative, ranking, rating or scaling, is presented in terms of the relative importance of the decision criteria used in the analysis. The result is a weighting checklist.

A Scale -Weighting Approach
Within this model, the importance weight for each criterion is multiplied by the scale, rating or ranking of each project option, and the resulting products in each case are then totalled to give an overall composite index or score for each option. This index can be written in the following form:

$$I_j = \sum_{i=1}^{n} w_i x_{ij} \tag{2-1}$$

where

I_j is the composited index for jth project option
n is the number of criteria
w_I is the importance weighting of the ith criterion
x_{ij} is the scale, ranking or rating of jth project option for the ith criterion

The information contained within a simple checklist is the minimum required for multi-criteria decision making to proceed. Data in this form is often assembled and presented at the outset of the process, and is used as a precursor to the development of information in the form of one of the other more complex checklists described above.

Examples of Checklists Used in Engineering Decision Making

Adkins and Burke (1974) devised a rating checklist for the comparison of two transportation route options in Texas. Each project option was assessed on the basis of a set of transportation, environmental, sociological and economic criteria. In each case the scale ranged from a minimum of -5 to a maximum of +5, with the maximum score indicating the most favourable outcome on the criterion under examination. Figure 2.1 below shows summary rating comparisons relative to all the major criterion headings.

Parameters	No. of plus ratings	No. of minus ratings	Total no. of ratings	Algebraic sum of ratings	Ratio of plus ratings	Average rating
Transport						
Option 1	15	6	21	52	0.714	2.476
Option 2	10	3	13	8	0.769	0.615
Environment						
Option 1	15	1	16	44	0.94	2.75
Option 2	12	2	14	14	0.86	1.00
Sociological						
Option 1	18	2	20	58	0.9	2.9
Option 2	12	4	16	6	0.75	0.375
Economic						
Option 1	15	14	29	27	0.52	0.93
Option 2	14	14	28	-11	0.50	-0.39
All Ratings						
Option 1	63	23	86	181	0.73	2.10
Option 2	48	23	71	17	0.68	0.24

Figure 2.1 – Comparison of ratings Using Adkins and Burke

The overall evaluation for each project option is based on the number of plus and minus ratings it received plus its algebraic average rating. At no stage are the ratings on different criteria for one option combined to give an overall score, thereby enabling direct comparison with the other.

Dee (1972) devised a scale-weighting checklist for evaluating water resources projects on the basis of mainly environmental criteria. It utilises a checklist of 78 environmental and socioeconomic parameters, each with their own predetermined importance weighting. For each option being considered, quantitative estimates of all decision criteria are normalised on a common scale ranging from zero to one. The final score for each parameter is determined by multiplying its normalised score by its importance weighting. The combined score for each option is calculated by summing the final score for each individual criterion:

$$E_j = \sum_{i=1}^{n} w_i u_{ij} \tag{2-2}$$

where

E_j = environmental impact units for jth option

u_i = environmental quality scale value for the *i*th criterion and the *j*th option
w_i = importance score for the *i*th criterion

Each option can then be directly compared with the others. This methodology, known as the Battelle System, has been used directly or in modified form within the environmental decision making process for many water-resources projects, including the Pa Mong water resources project in South Thailand (ESCAP, 1990). Its general approach has, however, been applied to other projects types, including a proposed rapid transit system (Smith, 1974), highway projects, pipeline projects and wastewater treatment installations (Dee, 1973).

An example of the use of scale-weighted checklists for the assessment of project options on a much broader range of criteria is the optimum-pathway matrix developed by Odum et al. (1971). This methodology was applied to the evaluation of eight project options for one section of an interstate highway in Georgia, USA. Initially, the analysis consisted of identifying the criteria to be included in the evaluation. A total of 56 criteria were identified, sorted into four main groupings as follows:

- economic and engineering / technical criteria,
- environmental and land-use criteria,
- recreation related criteria, and
- sociological criteria

Having compiled data for each of the 56 criteria, this information was then normalised for each option on each of the criteria, as follows:

$$S_i = \frac{1}{Max\{x_1, x_2, x_3, \ldots, x_8\}} \tag{2-3}$$

where

S_i is the scaling factor for the ith criterion, and $x_1, x_2, \ldots, x_8$ are the values of the eight project options on the ith criterion,

and

$$U_{ij} = S_i x_{ij} \tag{2-4}$$

Where U_{ij} is the normalised value for the ith criterion and the jth option, and x_{ij} is the assessment of the ith criterion for the jth option . The decision maker then determines a normalised set of importance weightings, w_i for each criterion.

The overall Battelle index for each project option I_j is estimated as follows:

$$I_j = \sum_{i=1}^{56} w_i S_i x_{ij} + e_i (w_i S_i x_{ij}) \tag{2-5}$$

where $e_i(\ )$ is an error term to allow for misjudgement on relative weightings of criteria and is assessed by computational analysis using a packaged stochastic computer program.

Comment

The scale / rank / rating-weighted checklist, which is the most complete form of the methodology, is an engineering application of the simple additive weighting (SAW) method, one of the best known and most widely used multi-criteria decision making methods.

The value / score of a given project option in the SAW method can be expressed mathematically as follows:

$$V_i = \sum_{j=1}^{n} w_j v_j (x_{ij}), \qquad i = 1, \ldots, m \tag{2-6}$$

Where V_i is the value function of the ith. project option, and w_j and $v_j\,(x_{ij})$ are weight and value functions of the jth.criterion respectively. The normalisation process allows direct addition among criteria to be achieved. The value of the ith option can then be expressed as:

$$V_i = \sum_{j=1}^{n} w_j r_{ij}, \qquad i = 1, \ldots, m \tag{2-7}$$

where r_{ij} which equals x_{ij} / S_{ij}, is a normalised x_{ij} (Yoon and Hwang, 1995).

The underlying assumption of the SAW method is that the decision criteria are preferentially independent. In the context of a checklist, this implies that, for a given project option, the contribution of an individual criterion to the overall score is independent of all other criterion scores. Within SAW also, there is a presumption that the weights are proportional to the relative value of a unit change in each criterion's value function.

2.3 Multi-Attribute Utility Methods (MAUT)

Multi-Attribute Utility Theory (MAUT) (Keeney and Raiffa, 1976) is a methodology which, within the context of an engineering project, allows possible consequences to be 'traded off' against one another, while also taking account of their probabilities of occurring. The closely related ideas of value and utility have a long history and are used in a wide variety of decision-making contexts. Engineers and planners use them when considering the best options for large-scale projects; especially those related to infrastructure development. Economists use them when analysing the operation of enterprises, markets and economies and especially in the field of welfare economics. Psychologists and social scientists use them in the study of how people behave and the reasons for the choices they make. Many of the concepts, assumptions and methods used are very similar in all these applications although some of the technical terms may be different or have different meanings in each subject area. The goal is to improve understanding of peoples' preferences, both as individuals and in groups, and to develop tools to assist in making decisions which correspond to these preferences. It is assumed that such decisions are good ones and that they will be accepted by a large number of the people affected by them.

In the eighteenth century Daniel Bernoulli considered the St. Petersburg paradox and the reasons for the success of lotteries and insurance schemes. He realised that a person's reaction to uncertainty when choosing between options was complex and did not necessarily coincide with simple measures of probability and benefit such as expected value. He understood that a person's "satisfaction" with a particular outcome was not necessarily in direct proportion to the magnitude of the outcome. For example the first glass of water gives much more satisfaction to a thirsty person than the second glass, even though both contain the same amount of water. He realised also that uncertainty and risk were key influences and he worked with the idea of expected utility.

INSURANCE People make regular small payments to insurance companies to cover themselves against loss caused by large catastrophes, which have small probabilities of occurring. Most insurance companies make money so it follows that its clients are willing to pay more than the expected values of the insured loss. Why?

LOTTERIES People make regular small payments to a lottery company to give themselves a very small probability of winning a very large amount of money. Most lottery companies make money so it follows that they pay out much less than they receive. Thus the expected value of the lottery must be less than the price of a ticket and yet many millions play lotteries. Why?

The classical situation is when the Decision-Maker must choose or recommend one or more option(s) from a set of available options. Each option, if selected, can lead to a number of consequences, but there is some uncertainty involved so that there is a certain probability that any given option will lead to any specified consequence. For the moment we consider only the case where there is only one type of consequence and we develop the ideas and principles of utility. Later on we show how the methods can be expanded to include many types of consequences in what is called Multi-Attribute Utility Theory (MAUT).

2.4 Utility

Utility is a concept which expresses a person or group's (called the Decision-Maker, DM) level of satisfaction with a particular outcome and is used here to determine preference or indifference between the outcomes/consequences of any set of options. The idea is a simple one and is based on the following assumptions about the Decision-Maker's preferences.

Existence
A Decision Maker, when considering any two of the options available will either prefer one to the other or be indifferent between them.

Transitivity
If option A is preferred to option B and option B is preferred to option C then option A is preferred to option C.

Monotonicity
The utility of any option is between the utility of any better option and any worse option. This means it can be expressed as a weighted average of the utility of any better and any worse option and that the weight lies between 0 and 1.

Probabilities Exist
The probabilities of each of the uncertain consequences occurring when any of the options are chosen exist, can be calculated and are known to the decision maker.

Monotonicity Of Probabilities
If two options can produce the same beneficial consequence, but with different probabilities then the Decision Maker will prefer the option which gives the higher probability of that consequence.

Substitution
If a Decision Maker has the same utility for two possible options, say A and B. then these can be substituted for one another in further decision making situation choice without changing the result.

Lotteries

The concept of a lottery is central to the application of utility. A lottery is a list of possible outcomes, say x_1, x_2, ,x_n, each having a probability of occurrence, p_1, p_2, ,p_n. A binary lottery has two possible outcomes, say x_1 and x_2, and x_1 has a probability, p, of occurring and x_2 a probability of (1-p). This is very often written as $(x_1,p: x_2)$. In utility theory lotteries can be compared and can be assigned utilities. In particular the **Substitution** property allows equivalent lotteries to be interchanged and also allows a lottery to be interchanged with a certain outcome.

2.5 Measurement of utility

Utility is expressed on an ordered metric scale. The numbers of this scale have no absolute physical meaning and the scale is constructed by assigning numbers to any two points. Usually these points correspond to the best possible outcome and the worst possible outcome for attribute in question. Very often the best possible outcome is assigned a utility of 1 and the worst a utility of 0. The basic principle use to construct the utility function from these two fixed points is that the utility of any binary lottery can be related to the utilities of its component outcomes and the probability of their occurrence.

For example, given option *a* yielding consequence x with a probability of p, and consequence y with a probability of 1-p, the utility of *a* can be written in terms of the utilities of x and y:

$$U(a) = pU(x) + (1 - p)U(y)$$

Utility functions not only involve the choice of the best option from a number of options on the basis of a family of relevant criteria, but also provide a framework for dealing with uncertainty. If the decision maker is asked to decide on the best option, given certain probability distributions for the criteria in question, he can state his preference between options A' and A" where

- Option A' results in criterion valuation x_i with a probability p'_i for $i = 1,n$, and
- Option A" will result in criterion valuation x_i with probability p''_i for $i = 1,n$.

A Utility function U has the property that, for a given criterion, say x_1, given two project options A and B each with a probability of giving a particular value of x_1 according to probability distributions $p_A(x_1)$ and $p_B(x_1)$, the project option with distribution A is preferred to the option with distribution B if:

$$E_A[U(x_1)] \geq E_B[U(x_1)] \tag{2-8}$$

where E_A, E_B are the expected utilities of A and B taken with respect to the distributions measures $p_A(x_1)$ and $p_B(x_1)$ respectively.

If two outcomes are identified by the Decision-Maker as equivalent and one of them contains an unknown quantity, either a probability or utility and the other probabilities and utilities are known, then the unknown can be determined from the equivalence. A Decision-Maker is usually interviewed and responds to a number of hypothetical situations designed to efficiently determine his utility function. Each response is used to define a point on the utility function. Two well known methods have been used:

Certainty Equivalent

Let L be the lottery yielding consequences $x_1, x_2, ..., x_n$ with probabilities $p_1, p_2, ..., p_n$ respectively. The expected value of the lottery, $\bar{x}$, is

$$\bar{x} = \sum_{i=1}^{n} p_i x_i \tag{2-9}$$

The expected Utility of this lottery is:

$$E[U(x)] = \sum_{i=1}^{n} p_i u_i(x_i) \tag{2-10}$$

A Certainty Equivalent of the lottery L is an amount x such that the decision maker is indifferent between L and the certainty of achieving consequence x.

The Certainty Equivalent is unique for any monotonic Utility function.

Certainty Equivalent Method of Forming the Utility Function

The extremes of the range of possible consequences are defined, e.g. x_{best} and x_{worst}. The worst one is assumed to have a utility of 0 and the best a utility of 1. The Decision-Maker is presented with a binary lottery generated from both extremes with a specified probability, p, of the best outcome occurring, i.e. (x_{best}, x_{worst}). Any probability, p, can be used but very often it is 0.5 The Decision-Maker is asked for the certain amount, x_1, equivalent to the lottery. Two new lotteries are generated by substituting this certainty equivalent for the best case and for the worst case and the Decision-Maker is asked for their certainty equivalents, x_2 and x_3. The process can be repeated with these new values until the utility function has been adequately defined,.

This method was widely used in the past but has a number of shortcomings.

- In practise, utility functions obtained using different values of p can be quite different.
- There is some concern about the appropriateness of comparing an uncertain outcome with a certain outcome.
- Since at each step, the utility is calculated using the utilities determined at previous steps, errors can propagate through the entire calculation.

Lottery Equivalent Method

As before, the extremes of the range of possible outcomes are determined and assigned utilities of 0 and 1. The Decision Maker is asked for the probability, p, which makes a lottery constructed from the extremes of the outcomes, i.e. $(x_{best}, p; x_{worst})$ equivalent to a lottery constructed from the worst extreme and some intermediate value and a specified probability, often 0.5, i.e. $(x_1, p; x_{worst})$.The utility of x_1 is then twice the resulting value of p.

In this method the comparison is always between two uncertain outcomes. The utility calculated at each step is independent of utilities determined at previous steps so errors do not propagate.

In both the certainty equivalent and the lottery equivalent methods, the interview process can be automated and computer programs have been written for this purpose. They can have some advantages; i.e. they are unbiased and very often people will interact more openly with a computer program than with an interviewer. This seems an interesting application for an expert system.

Multiattribute Utility Theory

In many cases the decision problems facing Engineers and Planners involve a large number of different attributes or types of consequences. In particular, decisions based on Environmental Impact Assessments may involve a very large number of types of consequences relating to water, air, noise, amenity, landscape, flora, fauna etc. In principle the same Utility Theory developed for the single decision attribute can be directly extended to cover such cases.

'Any decision-maker implicitly attempts to maximise some function 'U' which aggregates all the different view points to be taken account of'. In other words, if a decision maker is asked about his preferences from a range of options, his answer will be both coherent and consistent with a certain unknown function 'U'. This function 'U' will now be expressed in terms of a number of relevant attributes or criteria.

Estimating the form of this function is basic to the problem solving process within MAUT.

Direct Extension of Utility Theory

In principle the multi-attribute utility function can be measured by a direct extension of the way it is done for a single attribute utility function. The utility for two arbitrary reference points is defined and the utility for all other points can be estimated in relation to these. The different amounts of variation in the different consequences and in the combinations of consequences could be tested. In practise the amount of testing/surveying etc. required increases very quickly, proportional to the power of N, and the amount of data required becomes prohibitive, even for small numbers of attributes and especially for decisions with large numbers of environmental impacts. For instance if there are N attributes and each is to be defined by an empirical relationship fitted to, say P experimental points. Then the amount of cases to be tested is $P^N - 2$. Two of the points are fixed arbitrarily.

For example, suppose that a single attribute utility function could be adequately represented by 5 points. If there were two decision attributes then the utility function would be a two dimensional function and $5^2 - 2$ or twenty five points less the 2 fixed points would be required to represent the utility function with a corresponding level of accuracy. Thus twenty-three lottery equivalences must be established though questionning the decision maker. If there are three attributes then $5^3 - 2$ or one hundred and twenty three equivalences must be established if the utility function is to be represented with the same resolution. It is easily seen that the latter is quite impractical and would exhaust the patience of both analyst and decision maker. Even the two dimensional case requires considerable effort if tackled in the direct way. Fortunately this is not required and methods of constructing a multiattribute utility function without such extensive comparisons have been devised.

Keeney & Raiffa (1993) explored the conditions necessary to reduce the problem of "dimensionality" in the construction of the multiattribute utility function. A good introduction to the application of MAUT is given in De Neufville (1990)

Weighted Average

In the simplest approach, if the utilities of each type of consequence are independent of the other consequences then the multiattribute utility function is constructed as a weighted average of the utility functions for each individual attribute (consequence), i.e.

$$U(X) = \sum_{all\, i} w_i U(x_i) \qquad (2\text{-}11)$$

where $X = (x_1, x_2, \ldots, x_n)$ is the n-element vector of consequences (attributes),

Keeney & Raiffa (1993, p.286) express the above condition as "If the preferences over lotteries on x_1, x_2 and x_3 depend only on their marginal probability distributions and not on their joint probability distribution".

The w_i are weights which specify the relative contribution of each attribute in the final decision. In some situations these could be amounts of money, i.e. prices. There is an assumption here that such prices are constant regardless of the magnitude of the consequence and also are independent of the magnitude of any other consequence. This ignores any possible interaction between consequences. It is a perfectly valid approach as long as there are no such interactions. It has been used in many applications, cf. Vincke (1992) and is often called the additive model.

Multiplicative models

In many practical situations however, the utility of some consequences is influenced by other consequences and the simple weighted average approach cannot be used. For instance, the appreciation of visual amenity will depend somewhat on air quality. Keeney & Raiffa developed an approach for such cases which is based on two assumptions that, if valid, reduce considerably this problem of dimensionality.

PREFERENTIAL INDEPENDENCE: Suppose a decision problem has N different types of criteria. If for two of these a particular combination, say x_1^* and x_2^* is preferred to x_1^{**} and x_2^{**} then preferential independence requires that x_1^* and x_2^* will always be preferred to x_1^{**} and x_2^{**} for any values of the other consequences. This does not mean that the utilities remain the same in both cases since if the values of the other attributes change then the utilities may change. However it does mean that whatever the changes in utility the preference for the combination x_1^* and x_2^* does not change. A more formal definition of preferential independence is as follows:

The pair of criteria x_1 and x_2 is preferentially independent of x_3 if the conditional preferences in the (x_1, x_2) space, given x_3, do not depend on the value of x_3, i.e. if (x_1, x_2, x_3) is preferable to (x'_1, x'_2, x_3) for any given value of x_3 then it is preferable for all values of x_3.

For example, there may be several technical benefit criteria, and several technical cost criteria, and it may be that the conditional preferences among various packages of technical benefit levels may not depend on the technical cost levels involved.

Under conditions of certainty, preferential independence is the key concept in breaking a utility function down into its component parts which can then be summed up to give the final valuation. To deal with the uncertainty case, however, we must introduce and develop the concept of utility independence.

UTILITY INDEPENDENCE: Any indifference established for a particular consequence between a lottery and a certainty equivalent is not influenced by the other types of consequences.

When these two assumptions hold, the multiattribute utility function can be related to the individual utility functions by the equation

$$KU(X) + 1 = \prod_{i=1}^{n} \{1 + Kk_i u_i(x_i\}$$

where, both $U(X)$ and the $u_i(x_i)$ are scaled so that 0 represents the worst possible situation and 1 the best possible situation.

The multiplicative form of the equation allows a cross influence between consequences. This is best illustrated by expanding the equation for the case of three attributes. This gives

$$U(X) = k_1u_1(x_1) + k_2u_2(x_2) + k_3u_3(x_3) + K[k_1k_2u_1(x_1)u_2(x_2) + k_1k_3u_1(x_1)u_3(x_3)+ k_2k_3u_2(x_2)u_3(x_3)] + K^2 k_1k_2k_3u_1(x_1)u_2(x_2)u_3(x_3) \tag{2-12}$$

It is readily seen that this includes the simple weighted average as a special case, but also allows for all possible multiplicative combinations of cross influences between attributes.

The individual factors k_i must be determined as part of the analysis. They depend on the range of possible outcomes considered for each consequence and should not be interpreted as weights. They determine the value of K, i.e.

$$K + 1 = \prod_{i=1}^{n} \{1 + Kk_i\} \tag{2-13}$$

The multiattribute utility function can thus be constructed from N single-attribute utility functions and N values of k_i. This reduces considerably the investigation/surveys required. The steps involved are:

- specification of the attributes
- verification of assumptions
- measurement of the one-dimensional utility functions $u_i(x_i)$
- measurement of the scaling factors, k_i.
- calculation of the normalising parameter, K.
- generation of the multi-attribute utility (using equation above)

In the multiplicative form, when all the k_i values sum to one, K becomes equal to zero, and the equation reduces to the additive form.

$$U(X) = \sum_{i=1} nk_iU_i(x_i) \tag{2-14}$$

This occurs when all criteria are additively independent. It is a much stricter condition than that of preferential independence. Taking two criteria from an engineering decision problem, say noise pollution and air pollution, preferential independence implies that less noise pollution is preferable to more noise pollution, regardless of existing water pollution levels. Additive independence, on the other hand implies marginality between the two criteria i.e. their preference comparison depends only on their marginal probability distributions.

k will not equal zero when the sum of k_i values differs from 1. The sum can be less than or greater than 1 indicating that the criteria exhibit 'complimentary' or 'substitutive' interrelationships. In the case of a complimentary relationship, the combined effect of two criteria would be greater than their respective individual effects, and the sum of the scaling values k_i would be less than 1.0. Unlike the additive model, the use of which in this situation would be incorrect, the multiplicative model could account for effects which individually are not significant but cumulatively are influential. The opposite situation applies for the substitutive condition, where the scaling value sum is greater than 1.0.

For a given engineering decision problem, the scaling constants k_i are evaluated and then added up to determine the appropriate form of utility function to use.

2.6 Analytic Hierarchy Process (AHP)

Introduction

The Analytical Hierarchy Process (Saaty, 1980) is a multicriteria decision-aid methodology which allows qualitative data to be transformed into pairwise comparison data. It is essentially the formal expression of the decision maker's understanding of a complex problem using a hierarchical structure.

Saaty describes the theme of the Analytic Hierarchy Process (AHP) as a combination of 'decomposition by hierarchies and synthesis by finding relations through informed judgement'. Its purpose, as with most other decision-aid methods, is to develop a theory and provide a methodology for modelling unstructured decision choice problems in economic, social and engineering sciences. It reduces a decision problem to a series of smaller self-contained analyses. The relative merit of each project option is determined from a pairwise analysis of the relative performance ratings for all combinations of project options, separately for each decision criterion involved. The relative importance of each criterion is also determined from a similar pairwise analysis of decision makers preferences. The result of the overall process is a ranking of all options on an interval scale, enabling the optimal one to be selected.

AHP rejects the simplification of a problem's parameters in order to suit quantitative / monetary models such as cost - benefit analysis (CBA), believing it best to deal with complex situations as they are.

To understand more of the AHP method, we must define the nature of the hierarchies within it, and how the priorities within and between the hierarchies are established.

Hierarchies

A hierarchy enables the decision problem to be broken down into individual elements whose relationships with each other can then be analysed. Stated more formally by Saaty, it is 'an abstraction of the structure of a system to study the functional interactions of its components and their impacts on the entire system. The 'structure' and 'function' of a system cannot be separated. The former is the arrangement of its parts, and the latter is the function or duty which the components are meant to serve.

Thus, a hierarchical system is based on the assumption that the entities identified as relevant to the decision can be grouped into disjoint sets, with each set directly affecting the one above it. This is the 'decomposition' referred to in the opening definition. This hierarchical structure illustrates how results at the higher levels are directly influenced by those at the lower levels.

Hierarchies have many advantages. They can be used to describe how changes in priority at upper levels affect priorities of elements in lower levels. They provide detailed information on both the structure and function of the system, they are stable and flexible, and they accurately mirror reality, since natural systems are assembled hierarchically.

A hierarchy has at least three levels: the focus or overall goal of the decision problem at the top, multiple criteria that define alternative proposals in the middle layer, and competing options at the bottom. Take for example the formulation of the most appropriate transport strategy for a major urban centre. At Level 1, the focus would be on the desired end product of the process - a cost effective, technically efficient, socially equitable and environmentally sustainable transport strategy for the urban centre in question. Level 2 would comprise the criteria that are at the basis of the selection of the most appropriate strategy. These might include effect on regional employment rate, efficiency of implementation, effect on the overall quality of life of the urban population, consistency with European transport policies and coherence with local, regional and national development plans. Level 3 would consist of the various project strategy options: a major package of road investment, the construction of a new light-rail system for the area, an upgraded network of high quality bus corridors, an extension of the existing heavy rail urban network, plus an option which involves making better use of existing transport assets in the area.

Priority in Hierarchies

A hierarchy, by itself, is not a very powerful tool for decision making, unless the strength with which the various elements in one level influence those on the next higher level can be determined. In this way, the relative strengths of the impacts of the elements at the lowest level on the overall objectives can be computed. This is the 'synthesis' referred to in the opening definition.

The strengths of priorities of elements in one level relative to the next is determined as follows. Given the elements of one level, and one element X on the next higher level, the elements of the lower level are compared pairwise in their strength of influence on X. The numbers reflecting these comparisons are inserted in a matrix, and the eigenvector with the largest eigenvalue is found. The eigenvector itself provides the priority ordering, and the eigenvalue is a measure of the consistency of the judgement.

Saaty suggests using a simple nine point numerical scale, such as the one given in the table immediately below, to represent the results of each pairwise comparison. A nine-point scale is needed because Saaty (1977) believed that, within the framework of a simultaneous comparison, one does not need more than 9 scale points to distinguish between stimuli. Results from psychological studies (Miller, 1956) have shown that a scale of about 7 points was sufficiently discerning. Saaty noted that the ability to make qualitative decisions was well represented by five attributes (equality, weak preference, strong preference, very strong preference and absolute preference)

absolutely preferable	9
very strongly preferable	7
strongly preferable	5
mildly preferable	3
equal importance	1

For example, given four elements A,B,C,D within one hierarchy level, each pair - AB, AC, AD, BC, BD, and CD is directly compared with respect to their influence on X. If, for instance A is mildly preferable to B then the number 3 is placed in the cell at the intersection of the row corresponding to A with the column corresponding to B. Its reciprocal is placed in the symmetrically opposite cell. Adding all the possible pair-wise comparisons gives the matrix:

	A	B	C	D
A	1	$a_{1,2}$	$a_{1,3}$	$a_{1,4}$
B	$a_{2,1}$	1	$a_{2,3}$	$a_{2,4}$
C	$a_{3,1}$	$a_{3,2}$	1	$a_{3,4}$
D	$a_{4,1}$	$a_{4,2}$	$a_{4,3}$	1

Note that

$$a_{i,j} = \frac{1}{a_{j,i}} \tag{2-15}$$

Since an element is always equally important relative to itself (reflexivity), the main diagonal of the matrix always consists of 1's. The scores 2,4,6 and 8 and their reciprocals can be used as compromises between Saaty's categories.

An example of a numerated matrix could be as follows:

	A	B	C	D
A	1	5	6	7
B	1/5	1	4	6
C	1/6	1/4	1	4
D	1/7	1/6	¼	1

The next step involves the computation of a vector of priorities from the given matrix - known as the Eigenvector Method for Determining Weights.

The Eigenvector Method for Determining Weights

The eigenvector method for determining weights from pairwise comparisons is used in assessing the relative importance of n options. If the pairwise comparisons of their relative weights are known or estimated from surveys of relevant 'experts' or decision makers, the priority rating of options can be found by evaluating the priority vector or eigenvector of a matrix A whose elements are the pairwise comparisons of the weightings.

Assuming there are n criteria at some level of a decision tree, the process of pairwise comparison will require a total number of nC2 or n(n-1)/2 judgements. If the true weights for all criteria are known, then the result of a pairwise comparison, $a_{i,j}$ will simply be the ratio of the weight of criterion i to that of criterion j, i.e.

$$a_{i,j} = \frac{w_i}{w_j} \tag{2-16}$$

The reciprocal matrix is thus

	A	B	C	D
A	w_1/w_1	w_1/w_2	w_1/w_3	w_1/w_4
B	w_2/w_1	w_2/w_2	w_2/w_3	w_2/w_4
C	w_3/w_1	w_3/w_2	w_3/w_3	w_3/w_4
D	w_4/w_1	w_4/w_2	w_4/w_3	w_4/w_4

It is easily seen that the vector of weights itself $w^T = (w^1, w^2, \ldots\ldots, w^n)^T$ is an eigenvector of the reciprocal matrix. Once the PCM is constructed, the problem of determining the weights becomes that of finding an eigenvector W to satisfy the following equation:

$$Aw = \lambda_{max} w \tag{2-17}$$

where λ_{max} is the largest eigenvalue of the matrix A.

Relationship between AHP and MAUT

General

Although it is often assumed that the two approaches are competitors - AHP and MAUT, Belton (1986) presented the Analytic Hierarchy Process as a variation of the simple additive weighted value function utilised in multi-attribute value theory, i.e. Saaty's AHP model can be expressed in the following mathematical form:

$$V_i = \sum_j w_j x_{ij} \tag{2-18}$$

Where v_i is the overall valuation for option i, w_j is the weight assigned to criterion j, and $x_{i,j}$ is the score of option i on criterion j.

Within AHP, although weights w_j and scores $x_{i,j}$ are not explicitly distinguished as they are in MAUT, the weights are the result of the reconciliation of the judgement matrix or matrices of pairwise comparisons of the importance criteria, and the scores stem from pairwise comparisons of the options with respect to each criterion of the family of criteria being used for decision-making purposes.

Within MAUT, however, an explicit distinction between weights and scores is made, both in terms of theoretical definition and in the practical means of assessment. It uses a ratio scale of measurement for the criteria (distance between ratings is measurable, as opposed to just ranking numbers as in an ordinal scale) and an interval scale for the scores (minimum and maximum values are defined as zero and one). AHP expresses both weights and scores on the same 9 point ratio scale.

Score Assessment
In AHP, a pairwise comparison matrix is constructed using a 9-point semantic scale as referred to above - 'equally preferred', 'weakly preferred', 'strongly preferred' etc. AHP treats these semantic judgements as ones on a ratio scale. Even if one accepts that this link exists, the exact relationship between the semantic scale and the ratio scale is not defined, i.e. what is the distance between 'weak' and 'strong'?, and is it the same as the distance between 'very strong' and 'absolute'? Despite Saaty's insistence on the adequacy of the nine-point scale, Freeling (1983) believes that this scale imposes unnatural restrictions on judgements.

MAUT, on the other hand, assesses scores on an interval scale, requiring two arbitrary points at the end / extreme points of the scale to be specified. These anchor points represent a 'least tolerable' point - usually give a value 0, and a 'best attainable' point - usually given a value of 1 or 100. The scoring is thus more precise and defined than AHP, although judgements between two options using this interval scale can still lack reliability.

Weight Assessment
In AHP, the procedure for calculating weights is identical to that for evaluating scores, again involving the estimation of the eigenvector of weights from the matrix of pairwise comparisons of criteria.

The method adopted by the MAUT approach for the calculation of scores is, on the other hand, different to that used for weights, and usually involves the identification of the criterion which is weighted as most important, and the other criteria are then weighted relative to that one. The weights are usually normalised to sum to unity. A ratio scale of measurement is assumed.

Thus, while the distinction between scores and weights in MAUT is defined and clear, the distinction between the two in AHP is unclear, with the exact concept of weight remaining ambiguous and indistinct in nature from the procedure for scoring options.

Summary

Saaty's eigenvector approach to pairwise comparisons provides a framework for calibrating a numerical scale, and is useful in engineering problems where quantitative valuations and comparisons do not exist in every instance. The participation of a group of decision makers make it possible to assess trade-offs between diverse criteria or options at a given level of analysis. It can be used as an aid to the decision making process and enables compromise to be attained from the various judgements reached.

However, while AHP is a coherent easily understood method of evaluation, it's blurring of the concepts of scores and weights, and it's use of a limited semantic scale, which Saaty assumes can be directly related to a ratio scale, makes it less precise and comprehensive than the simple additive weighted model from which it is derived, and to which it can be related.

2.7 Concordance Analysis

General

Concordance Analysis is a non-compensatory Multi-Criteria Decision Making (MCDM) model. The method uses various mathematical functions to indicate the degree of dominance of one option or group of options (Massam, 1988). The model permits any two options to remain incomparable with each other. For example, if, in the context of a choice problem, some option a is better than both b and c, it becomes irrelevant to analyse preferences between b and c - they can remain uncompared without endangering the decision-aid procedure. Indeed, a conclusion of incomparability between some actions may be quite helpful, since it highlights some aspects of the problem which would perhaps deserve a more thorough study. Within Concordance Analysis, there is no question of the 'trading-off' of one criterion directly against another for each individual option. In contrast, compensatory/additive models such as MAUT seek to reduce option evaluation and selection to one in which each of the alternative options is classified using a single score which represents the attractiveness or utility of an option. The selection of a preferred plan is based upon these scores. They rely on the principle that trade-offs between criteria are legitimate so that the method seeks to identify these, and then determines a utility score for each option (Rogers and Bruen, 1995).

Comparison between options proceeds on a pairwise basis with respect to each criterion, and establishes the degree of dominance that one option has over another. Mathematical functions indicate this degree of dominance, determining the extent to which project outcomes and preference weights confirm or contradict the pairwise dominance relationships between alternative projects.

The method examines both the degree to which

1. the preference weights are in agreement with pairwise dominance relationships, and
2. the degree to which weighted evaluations differ from each other (Hwang and Yoon,1981).

These stages are based on what are defined as a concordance and discordance set. This twofold approach has the advantage that the available information is used as intensively as possible.

The result of the process is the selection of an optimum option or group of preferred options.

One of the most commonly used methods within Concordance Analysis, the ELECTRE Method, (Elimination et choix traduisant la realite) was originally developed by Benayoun, Roy et al. (1966). ELECTRE involves a systematic analysis of the relationship between all possible pairings of the different options, based on each option's scores on a set of common criteria of evaluation. The result is a measure of what is termed the 'outranking' of one option over another. Option *a* is said to outrank *b* if *a* is adjudged to be at least as good as *b*. Hence ELECTRE and other similarly based techniques are also known as Outranking Methods. While ELECTRE has no axiomatic basis, and incorporates the role of intuition and professional judgement, it nonetheless provides a valuable framework within which to examine multi-criteria problems.

The general characteristic of a Concordance Analysis is that it attempts to select a 'best option' out of a series of competing options on the basis of multiple criteria. All options are scored on each of the given criteria. The outcomes or scores for all options with respect to all relevant criteria can be included in an impact matrix P as follows:

$$P = \begin{bmatrix} p_{11} & \cdot & \cdot & \cdot & p_{1l} \\ \cdot & \cdot & \cdot & \cdot & \cdot \\ \cdot & \cdot & \cdot & \cdot & \cdot \\ \cdot & \cdot & \cdot & \cdot & \cdot \\ p_{j1} & \cdot & \cdot & \cdot & p_{jl} \end{bmatrix} \tag{2-19}$$

where a typical element $p_{i,j}$ of P represents the evaluation of the jth criterion for the ith option.

All scores on a certain criterion can be represented in any appropriate unit of measurement: a monetary, or even quantitative measurement is not necessary. The impact matrix P is only a technical-physical representation of the various individual criterion scores for each option No common evaluation of the options on the basis of the combined set of criteria takes place. The only condition for constructing the impact matrix is that the decision maker must be able to rank the outcomes for each individual criterion by means of an implicit preference criterion stating, for example, that a higher score is more desirable than a lower one.

The selection process requires that a set of preference scores be given for each criterion. In addition, a set of weights, W, must be in place, representing the relative preference weightings attached by the decision maker to each of the criteria. Thus, each criterion is assigned a weight w_j, (j = 1,n, n = number of criteria). The assignment of these importance weightings is intrinsic to the proper use of the method.

The most common way of determining a set of weights is by means of interview techniques. Using this format, a decision maker is asked to specify on an appropriate scale his relative priorities for a set of decision criteria. On the basis of the responses, a set of weightings is derived which reflect the decision maker's perception regarding the relative importance of the individual criteria. In the context of ELECTRE, a number of methods for weighting criteria have been devised (Mousseau, 1995, Simos, 1990, Hokkanen and Salminen, 1994, Rogers and Bruen, 1998). These are discussed in Chapter 3.

Given the impact matrix P and the weighting vector W, Concordance Analysis proceeds to select the best project option by means of direct pairwise comparison. The relative pairwise outranking of options can be determined through the definition of a concordance and a discordance measure for each pair separately. A complimentary analysis can then be used to eliminate the inferior, less well performing options, and then identify / choose the best performing option or group of options. In order to achieve this, the complimentary analysis will define threshold values for the concordance and discordance measures, from which definitive conclusions can be drawn regarding the rank ordering of the options under consideration. This is the approach adopted within the basic ELECTRE Method. This general methodology will be detailed immediately below. Details of the modern variants of the model are discussed in detail in chapter 3.

The Basic Concepts of the ELECTRE Methodology

This method is based on the direct pairwise comparison of all options under consideration. As the above definition of the method's name indicates, the twin processes of elimination and choice are essential components of ELECTRE. Firstly, the method attempts to eliminate a subset of less desirable options from the complete set of options, and, subsequently, it uses a complimentary analysis to select the best option or group of options.

Initially, a concordance set of criteria for each set of options are defined. For a given pair of options 1 and 2, the set of decision criteria $J = \{j \mid j = 1,2,\ldots,n\}$ is divided up into two subsets. The first, the concordance set, is composed of all criteria where option 1 is either preferred or judged equal to option 2. The second complimentary subset is called the discordance set. It represents the subset of all criteria for which option 1 is worse than option 2.

The concordance set will increase in size as option 1 outranks option 2 for successively more decision criteria. The relative value of this concordance set is estimated using the concordance index, which is equal to the sum of the weights of those criteria contained within the concordance set.

The concordance index for a pair of options 1 and 2 is the sum of the weights for the criteria for which options 1 is not worse than option 2, normalised by dividing by the sum of all the weights.

The index gives a set of values for $c_{1,2}$ which range from 1.0 where option 1 is equal to or better than option 2 for all the criteria, to 0.0 where option 1 is worse than option 2 for all the criteria.

The complete set of concordance indices are assembled in the concordance matrix C as follows:

$$C = \begin{bmatrix} c_{1,1} & c_{1,2} & \cdot & \cdot & c_{1,m} \\ c_{2,1} & c_{2,2} & \cdot & \cdot & \cdot \\ \cdot & \cdot & \cdot & \cdot & \cdot \\ \cdot & \cdot & \cdot & \cdot & \cdot \\ c_{m,1} & \cdot & \cdot & \cdot & c_{m,m} \end{bmatrix} \tag{2-20}$$

The concordance index indicates the relative dominance of one option over another, based on the relative importance weightings of the relevant decision criteria.

The matrix C is, in general, not symmetric.

Again, for a given pair of options 1 and 2, the discordance index measures the degree to which option 1 is worse than option 2.

The discordance index is the maximum value, over all the criteria, of the difference between the assessments for each pair of options divided by the maximum possible difference. It indicates the maximum of the differences between the pairwise scores over all the criteria belonging to the discordance set. In order that all values of the discordance index fall between zero and 1, a normalisation procedure is carried out by dividing the index by the maximum overall difference for the criterion in question. It is necessitated by the potentially differing dimensions of the criterion scoring systems.

c_{12} uses information on the weights for the criteria. Such information is not part of the calculation of d_{12}. If only ordinal rankings exist for each plan/option, only c_{12} indicators should be calculated. The d_{12} indicators require ratio data i.e. the distance between rankings must be measurable. Furthermore, if raw data / raw scores for one criterion for a set of options is standardised, and these values used, then the discordance index can depend on the standardisation procedure which is adopted, e.g.

$$1. \frac{Raw\ score}{\sum All\ scores} \tag{2-21}$$

$$2. \frac{Raw\,score}{Max\,score} \tag{2-22}$$

The complete set of discordance indices can be included in the (m×m) discordance matrix D as follows:

$$D = \begin{bmatrix} d_{11} & d_{12} & . & . & . & d_{1m} \\ d_{21} & . & & & & . \\ . & & & . & & . \\ . & & & & . & . \\ . & & & & . & . \\ d_{m1} & . & . & . & . & d_{mm} \end{bmatrix} \tag{2-23}$$

The information contained in D is complimentary to, and differs fundamentally from, that contained in C. Differences among criterion importance weightings are represented by the concordance index, whereas differences among criterion valuations are represented by the discordance index.

Concordance and discordance indices can be viewed as measurements of satisfaction and dissatisfaction that a decision maker feels on choosing one option over the other.

The best option is selected by considering the combination of Concordance and Discordance values for each pair of options. In the simplest / most trivial case in which one option, say option 1, is preferred to all others for all criteria, then $c_{1i'} = 1.0$ and $d_{1i'} = 0.0$ (i' referring to all other options). In this situation, obviously, option 1 outranks all others, and is thus the best option. In practice, however, an option with ideal concordance and discordance indices rarely exists.

Thus, once an option is outranked by one of the other options, it is *eliminated* from the process, as it is unlikely to be chosen as a preferred option. It is unlikely, however, that there will be one sole option which is non-dominated. Usually multiple non-dominated solutions exist. Each member of this *kernel* of preferred options may not be outranked by any of the other options either inside or outside the kernel. Also, each of the *eliminated* options outside the kernel must be outranked by at least one of the options within the kernel (Nijkamp and Van Delft, 1977). If the process results in multiple non-dominated solutions, and the decision maker requires one option possessing 'absolute dominance' to be identified, this may be achieved through altering the threshold values $\hat{c}$ and $\hat{d}$ so that stronger preference relationships are imposed. Revised threshold values for the two indicators are introduced in order to find one of the options which can be defined as the best. By successive trials, and by diminishing the strictness thresholds from $\hat{c} = 1$ (the optimum concordance value) downwards, and from $\hat{d} = 0.0$ (the optimum discordance value) upwards, one finally selects a best option through a process of elimination. As the strictness levels are relaxed, the outranking relationships derived are no longer unequivocal. The relationship implies that, given both the information known regarding the decision maker's preferences *and* the quality of the criterion valuations for each option, there are sufficient arguments to support the statement that option i is at least as good as i' (*concordance*), while no essential reasons exist to refute this (*absence of discordance*) (Vincke, 1992). (It must be borne in mind that this outranking relationship only refers to the dominance of i by i'. It has no bearing on the outranking relationship between i and any other option under consideration). Further differentiation between options within the kernel now becomes possible, allowing the kernel to be reduced in number / size. At some point, there comes a value of $\hat{c}$ and $\hat{d}$ at which only one option is retained within the kernel - this is the favoured solution. However, as Guigou (1971) pointed out, if this occurs at a value of $\hat{c}$ less than 0.6, for whatever value of $\hat{d}$, the relation will lose some of its realism. It would thus be difficult to give much significance to the result, i.e. if $\hat{c} = 0.6$ for c_{12}, it means that there is no more than a 60% likelihood that option 1 is at least as good as option 2. Thus the outranking relation becomes weaker and the ultimate choice does not appear as convincing and decisive. This procedure has been described by Van Delft and Nijkamp (1977) as an arbitrary, ambiguous and not very satisfactory approach. This shortcoming has been addressed in a number of the ELECTRE Models developed later by Roy, where various 'exploitation' procedures have been employed to rank those options within the kernel of non-dominated solutions. These models are detailed in Chapter 3.

This approach does not give an overall ranking of options which would allow the relative attractiveness of each to be judged. It initially results in the creation of a sub-set or 'kernel' of options, one of which will be deemed most desirable.

This method is stated in mathematical terms as Roy's ELECTRE I method within Chapter 3.

2.8 Suitability of models for use with environmental decision problems

The multi-dimensional nature of the four different multi-criteria decision models discussed in this chapter is one of their major advantages over the traditional monetary evaluation methods and their derivatives. On a general level, the two-staged nature of the analysis within Concordance Techniques gives it a basic advantage over the other three. Within Concordance Analysis, the information contained within the concordance matrix differs fundamentally from that contained in the discordance matrix. The information from the two sources is complimentary. Differences between weights are interpreted through the concordance matrix, while differences between criterion scores for all pairs of options are interpreted through the discordance matrix. The result is a potentially more refined analysis, and is an advantage compared with the other multi-criteria models where no such refinements exist.

In the context of their suitability in assessing engineering and infrastructure development projects, all four methods are, in general terms, appropriate. All can illustrate how different project options affect different groups within society, and can use and adjust weightings as part of a sensitivity analysis and to produce consistent results. They also have the capacity to assess a large number of decision variables.

Checklist Methods and Multi-Attribute Utility Methods (MAUT) provide a systematic, mathematically rigorous method for evaluating the overall effects of engineering projects. However, the limited quality of the data obtainable for such a wide range of criteria considered at the planning stage of a project may be such that the mathematical rigour of the method may obscure the uncertainties in the final result. Furthermore, the practice of 'trading-off' diverse impacts, which is central to these methods, could be seen as misplaced, given the potential diversity of criteria in terms both of their origin and data resource requirements.

AHP is also a coherent and systematic method. It possesses a simplicity which Index Methods lack, thereby making it more widely applicable. However, it blurs the concepts of scores and weights, which does not occur in Index Methods where it is designated that the former reflects the absolute valuation of an option on a given criterion, while the latter is taken to reflect the relative importance of a criterion to all others under consideration. In addition, it's use of a limited relatively coarse 9-point semantic scale limits it's ability to distinguish between differing valuations of the more sensitive criteria used within the analysis of complex engineering project, for example, those related to environmental impact.

Concordance Analysis / ELECTRE possesses some significant advantages over the other three methodologies. It is a rigorous yet adaptable method, allowing the evaluation of mixed qualitative and quantitative data. It's basis in the non-tradability of criteria, providing a sub-set of non-dominant options in many situations, may not provide a strict pre-order of options, but may reflect a more realistic solution to the problem of option choice given the quality of data available. Also, the detailed data requirements for the other three methods, requiring a relatively high level of resources in terms of time and manpower, may preclude their use within the assessment of an engineering project at the planning stage. Concordance Analysis / ELECTRE does not require resources at such a high level. Given the number and diversity of criteria potentially relevant to major engineering / infrastructure projects, this factor is important in the choice of method.

Recognising that the final choice of method will depend, to a large extent, on the type and quality of data obtainable within the resources available, Concordance Analysis / ELECTRE is recommended as the decision support method likely to be most appropriate for the assessment of complex Engineering and Infrastructure projects at the planning stage.

2.9 References

Adkins, W.G. and Burke, D. (1974) 'Social, Economic and Environmental Factors in Highway Decision Making'. *Research Report 148-4 Texas Transportation Institute*, Texas A&M University, College Station, November.

Belton, V. (1986) 'A Comparison of the Analytic Hierarchy Process and a Simple Multi-Attribute Value Function'. *European Journal of Operational Research*, Vol. 26, pp7-21.

Benayoun, R., Roy, B. and Sussman, N. (1966) 'Manual de Referance du Programme ELECTRE'. Note de Synthese et Formation, No. 25, Direction Scientifique SEMA, Paris.

Canter, L.W. (1996) *Environmental Impact Assessment*. McGraw Hill.

Dee, N.(1972) Environmental *Evaluation System for Water resources Planning*. Final Report, Battelle-Columbus Laboratories, Columbus, Ohio.

Dee, N.J. (1973) 'Environmental Evaluation System for Water Resources Research'. *Water Resources Research*, Vol. 9, pp523-535.

Economic and Social Commission for Asia and the Pacific (1990) Environmental Impact Assessment – Guidelines for water resources Development. ST/ESCAP/786, United nations, New York, pp19-48.

Freeling, A.N.S. (1983) *Belief and decision Making*. PhD Thesis, Cambridge.

Guigou, J.L. (1971) 'On French Location Models for Production Units'. *Urban Economics*, Vol. 1(2), pp107-138.

Hokkanen, J. and Salminen, S. (1994) Choice of a Solid Waste Management System by Using the ELECTRE III Method. *Applying MCDA for Decision to Environmental Management*. (Ed. M. Paruccini), Kluwer Academic Publishers, Dordrecht, Holland..

Hwang, C.L. and Yoon, M.J. (1981) Multiple Attribute Decision Making: Methods and Applications. Springer-Verlag.

Keeney, R.L. and Raiffa, H. (1976) Decisions with Multiple Objectives: Preferences and Value Trade-Offs. Wiley

Massam, B. (1988) *Spatial Search*. Pergamon Press.

Miller, G.A. (1956) 'The Magical Number Seven, Plus or Minus Two: Some Limits On Our Capacity For Processing Information'. The Psychological Review, Vol. 63, pp81-97.

Mousseau, V. (1995) Eliciting information concerning the relative importance of criteria. *Advances in Multicriteria analysis* (Pardalos, Y., Siskos, C. and Zopounidi, C. (eds.)), pp17-43. Kluwer Academic Publishers.

Odum, E.P. (1971) 'Optimum Pathway Matrix Analysis Approach to the Environmental Decision Making Process – Test Case: Relative Impact of Proposed Highway Alternates' Institute of Ecology, University of Georgia, Athens.

Rogers, M.G., and Bruen, M.P. (1995) 'Non-monetary Decision-Aid techniques in EIA - An Overview'. *Paper 10746, Proceedings of the Institution of Civil Engineers*, Municipal Engineer, 109, June, pp88-103.

Rogers, M.G., and Bruen, M.P. (1998) 'A New System for Weighting Criteria within Electre'. *European Journal of Operational Research*, Vol. 107, No. 3, pp. 552-563.

Saaty, T.L. (1980) *The Analytic Hierarchy Process*. McGraw-Hill.

Saaty, T.L. (1977) 'A Scaling for Priorities in Hierarchical Structures'. *Journal of Mathematical Psychology*, Vol. 15, pp207-218.

Simos, J. (1990) *Evaluer L'Impact sur L'Environment* Presses Polytechniques et Universitaires Romandes, Lausanne.

Smith, M.A.(1974) 'Field Test of an Environmental Impact Assessment Methodology'. *Report ERC-1574 Environmental Resources Centre*, Georgia Institute of Technology, Atlanta.

Vincke, P. (1992) *Multicriteria Decision Aid*. John Wiley.Van Delft and Nijkamp, 1977

3 THE ELECTRE METHODOLOGY

3.1 Introduction to ELECTRE

In July 1966, Bernard Roy presented a paper in Rome in which he used his formal training in Mathematics to develop a practical decision-making system, now known as ELECTRE. It involves a systematic analysis of the relationship between all possible pairings of the different options, based on each option's scores on a set of common criteria of evaluation. The result is a measure of the degree to which each option outranks all others. The methodology entails the construction of an outranking relation, the generation of concordance and discordance indices (including the notion of relative importance of each criterion) and an analysis of the results obtained from an overall evaluation of all the outranking relationships derived.

There are six main versions of ELECTRE - I, II, III, IV, Tri and IS. These are explained in detail in Sections 3.4 to 3.9. In each case, the project options are assessed in terms of several criteria, each criterion being a measure of the decision-makers preferences according to some point of view. The version of the ELECTRE model employed depends on the types of criteria involved. The definition of these criteria is thus of central importance to an understanding of ELECTRE.

3.2 Types of Criterion used with ELECTRE

The following four types of criterion are used with Multicriteria Decision Aid methods:

- a true criterion
- a semi-criterion,
- an interval criterion, and
- a pseudo-criterion

Of these, only the first and last types are used within ELECTRE. However, it is useful to understand the differences between all four.

True Criteria

These are the simplest form of criterion, used within what is termed a 'traditional' preference structure where no thresholds exist, and the differences between criterion scores are used to determine which option is preferred. The resulting ranking structure is known as a complete preorder.

Any preference or outranking structure can be completely characterised by the outranking relationship, **S,** which defines the conditions necessary for one option, a, to outrank another, b. Here option a outranks b if the decision maker prefers it to b or is indifferent between the two. This is written formally as

$$aSb \text{ iff } aPb \text{ or } aIb$$

Within the 'traditional' structure, the decision-makers preferences satisfy the following model:

$$aPb \Leftrightarrow g(a) > g(b)$$
$$aIb \Leftrightarrow g(a) = g(b)$$
$$\forall\ a,b \in A$$
$$\text{since } S = P \cup I,\ aSb \Leftrightarrow g(a) \geq g(b)$$

The indifference relation **I** is transitive, i.e.

$$aIb \text{ and } bIc \Leftrightarrow aIc$$

This preference structure is called a **complete preorder**. All options can be ranked from best to worst, allowing a tie between options of equal ranking. (If there were no ties, the relationship would become a **complete order**.)

True criteria are used in the ELECTRE I and II versions of the model.

Semi-Criteria

These are used within what is termed a 'threshold' model or preference structure, where a constant 'just-noticeable difference' (JND) exists for a given criterion. The difference between the scores for any two options must exceed this threshold JND before one option is declared superior to another. This allows account to be taken of possible errors and/or uncertainties in predicting the criterion valuations since the JND can be related to the uncertainties in the criterion evaluation or perception.

In this situation, the underlying preference structure is called a semi-order structure and unlike the traditional model it is intransitive. For example, when comparing noise levels between three sites X, Y and Z, with a threshold difference defined at 5dB(A), if X=70dB(A), Y=74dB(A) and Z=78dB(A), while X remains indifferent to Y and Y remains indifferent to Z, from the listeners point of view, the listener is not indifferent between the noise levels at X and Z (Z minus X is 8dB(A), which is greater than the defined threshold difference of 5dB(A).) Thus, within the threshold model, the Indifference relation **I** is not, by definition, transitive.

Thus, introducing a positive threshold q, $\forall a,b \in A$

$$aPb \Leftrightarrow g(a) > g(b) + q$$
$$aIb \Leftrightarrow |g(a) - g(b)| \leq q$$

These two relations define the **Threshold model**. A preference structure is thus a semi-order structure if it can be represented by a threshold model.

Interval Criteria

These are used within what is termed a 'variable threshold' model. In this case, the thresholds can vary with the scale of the criterion valuations being compared, i.e.

$$aPb \Leftrightarrow g(a) > g(b) + q(g(b))$$
$$aIb \Leftrightarrow g(a) \leq g(b) + q(g(b)) \text{ and } g(b) \leq g(a) + q(g(a))$$

These relationships define the **Variable Threshold Model.**

Pseudo-Criteria

Pseudo-criteria involve a two-tiered threshold approach. The variable threshold model defined above may seem unrealistic because it defines a precise value above which there is a strict preference, and below which there is indifference. Real life examples demonstrate that there is often an intermediary zone inside which the decision maker's information is contradictory or indeterminate. This has led to a preference model which explicitly includes two different thresholds, firstly, an indifference threshold q, beneath which the decision maker shows clear indifference, and, secondly, a preference threshold p, above which the decision maker is certain of strict preference. In between there are situations in which a “weak preference” for option a over b , denoted aQb applies, as follows;

$$aPb \Leftrightarrow g(a) > g(b) + p(g(b))$$
$$aQb \Leftrightarrow g(b) + p(g(b)) \geq g(a) > g(b) + q(g(b))$$
$$aIb \Leftrightarrow g(b) + q(g(b)) \geq g(a) \text{ and } g(a) + q(g(a)) \geq g(b)$$

Weak preference indicates the decision maker's hesitation between Indifference(**I**) and Strict Preference(**P**).

ELECTRE IS, III, IV and Tri use Pseudo-criteria.

3.3 Basic Concepts of ELECTRE

In general, within a multicriteria decision problem, Option *a* is said to outrank Option *b* if, given both the level of knowledge regarding the decision maker's preferences and the quality of information on all relevant criteria available for each option, sufficient arguments exist in favour of deciding that Option *a* is at least as good as Option *b*, and no compelling arguments exist to the contrary.

To construct an outranking relation, this definition must be enriched in such a way as to facilitate the solution of the decision problem. The ELECTRE method attempts this 'enrichment' in two distinct stages:

- the construction of the outranking relation, and
- the exploitation of this relation.

Each of these stages may be treated in a number of ways, depending on the problem formulation and the particular version of ELECTRE being used, as described in the following sections.

3.4 ELECTRE I

General

ELECTRE I was the first version of the model devised by Roy (1968) to solve multi-criteria choice problems. Its aim is to obtain a subset or kernel N of project options such that any action which is not in N is outranked by at least one action in N. It is not necessarily a set of preferred options, but is the set in which the best compromise can certainly be found.

The two phases of ELECTRE I can be described as follows:

Construction of the outranking relation

Assuming each criterion is assigned a weight w_j ($j = 1,n$; n = number of criteria), increasing with the importance of the criterion, the concordance index for each ordered pair (a,b) can be defined as follows:

$$C(a,b) = \frac{1}{W} \sum_{\forall j: g_j(a) \geq g_j(b)} w_j \qquad (3\text{-}1)$$

where

$$W = \sum_{j=1}^{n} w_j \qquad (3\text{-}2)$$

and $g_j(a)$ is the score for option a under criterion j.

C(a,b) has a value between 0 and 1, and measures the strength of the statement 'Option a outranks Option b'. However, any outranking of 'b' by 'a' which results can be weakened or negated by the Discordance index D(a,b), defined as follows :

$$D(a,b) = 0 \text{ if } g_j(a) \geq g_j(b) \ \forall j \tag{3-3}$$

Otherwise,

$$D(a,b) = \frac{1}{\delta} \max_j \left[g_j(b) - g_j(a) \right] \tag{3-4}$$

$$\delta = \max_{c,d,j} \left| g_j(c) - g_j(d) \right| \tag{3-5}$$

(δ = maximum difference on any criterion)

D(a,b) is thus an index, with its value between 1 and zero, which increases if the preference for 'b' over 'a' is very large for at least one criterion. It can only be used if scores for different criteria are comparable and thus not with qualitative data. If the discordance index D(a,b) reaches a certain threshold value, any outranking of 'b' by 'a' which might be indicated by the Concordance index, is prevented.

(Alternatively, the discordance index can be defined as follows:

$$D(a,b) = \max_j \left[\frac{g_j(b) - g_j(a)}{\delta_j} \right] \tag{3-6}$$

where δ_j is the range of the scale associated with the criterion j.

The practical consequence of this latter definition of D(a,b) is that the values of the discordance matrix tend to be greater than those obtained using the first definition of discordance referred to above.)

The outranking relationship for ELECTRE I is built by comparing the concordance and discordance indices with specified limits (thresholds). Suppose that limiting values have been specified for $\hat{c}$ (the concordance threshold -maximum of 1) and $\hat{d}$ (the discordance threshold - a minimum of 0). Then an outranking relation S is defined by:

$$aSb \text{ iff } C(a,b) \geq \hat{c} \text{ and } D(a,b) \leq \hat{d}$$

Exploitation of the Ranking

ELECTRE I, seeks to obtain a partition of the set of all options (A) into two subsets N and A \ N such that :

1. each option in A \ N is outranked by at least one option in N,

   i.e. $\forall\ b \in A \setminus N, \exists a \in N: aSb$

2. the option in N are incomparable with the defined outranking relationship, i.e. $\forall a,b \in N,\ a \not S b,\ b \not S a$.

In graph theory, the set N is called a kernel of the graph. If the graph has no circuit, the kernel exists and is unique. In any given case, the number of options in the kernel may decreases as the value of $\hat{c}$ is decreased from 1 and the value of $\hat{d}$ is increased from 0.

An Example of a Graph and its Kernel

The relationships built up by ELECTRE I can be readily expressed in the form of a graph. An illustration of this type of graph is given in Figure 3.1. Within it, each node is represented by a circled number. Each node corresponds to an option. In this case there are 4 options in the set. The arrows emanating from the nodes are called directed paths and correspond to an outranking relation. Here 4 outranks 1, 2 and 3, and 1 outranks 2. The kernel of the graph thus consists of alternative 4, and is the subset of non-dominated alternatives which the ELECTRE I method defines.

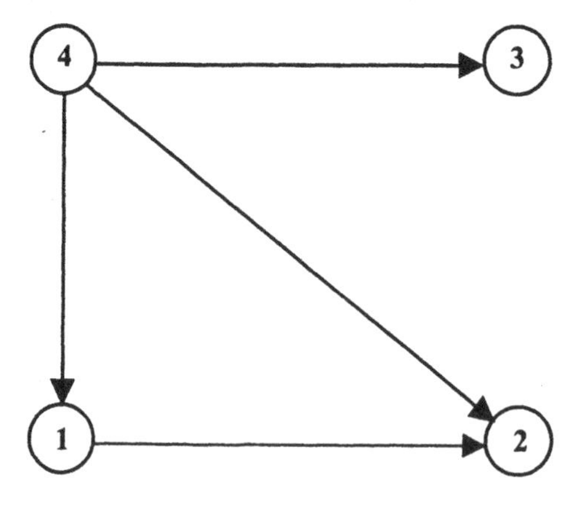

Figure 3-1

Sensitivity Testing for ELECTRE I

A sensitivity analysis tests the robustness of the results obtained from the decision model by varying the values of the weights and thresholds and observing the effect on the final outcome. The range of parameter values over which the result remains unchanged is indicated, and those variables crucial to a change in option selection can be highlighted. On the basis of such an analysis, it is possible to overcome some of the reservations not only of the decision-makers but also of the technical experts from whom the initial parameter values originated. If, in varying the parameters either side of their initial values, the results are not modified to any significant extent, they are robust.

The sensitivity tests in ELECTRE I, usually varies the values of the following:

- the size or range of the scales used to assess the criteria
- the criterion weightings (w_j)
- the concordance threshold ($\hat{c}$)
- the discordance threshold ($\hat{d}$)

Example - Solving an Outranking Problem using ELECTRE I

Suppose there are six alternative schemes (P1 to P6) for a proposed motorway project. Each option is assessed in terms of the following five environmental criteria:

Cr1 - noise pollution effects

Cr2 - community and land severance effects

Cr3 - air pollution effects

Cr4 - land use effects

Cr5 - recreation effects

Each criterion has the following importance weighting:

Criterion	**Cr1**	**Cr2**	**Cr3**	**Cr4**	**Cr4**
Weighting	**3**	**2**	**3**	**1**	**1**

Figure 3-2 - Criterion Weightings

Each criterion is assessed on the following qualitative scale:

Very Beneficial	**VB**
Beneficial	**B**
Neutral	**N**
Adverse	**A**
Severely Adverse	**SA**

Figure 3-3 -Rating Scale

The assessment of each impact/criterion, for each option, is as follows:

	Cr1	**Cr2**	**Cr3**	**Cr4**	**Cr5**
P1	N	VB	A	N	VB
P2	SA	A	A	VB	N
P3	SA	N	SA	VB	A
P4	VB	A	N	N	N
P5	VB	N	B	N	B
P6	VB	N	VB	B	B
Weighting	3	2	3	1	1

Figure 3-4 - Criterion Grades

For the purposes of this problem, the following scoring system, by Guigou (1971), is employed:

a) for Cr1,Cr2,Cr3 - VB=20, B=15, N=10, A=5, SA=0

b) for Cr4, Cr5 - VB=16, B=13, N=10, A=7, SA=4

This scoring system leads to a second table of unweighted information, shown below in Figure 3-5.

	Cr1	**Cr2**	**Cr3**	**Cr4**	**Cr5**
P1	10	20	5	10	16
P2	0	5	5	16	10
P3	0	10	0	16	7
P4	20	5	10	10	13
P5	20	10	15	10	13
P6	20	10	20	13	13

Figure 3-5 - Criterion Scores

Sample calculations for some of the concordance indices are as follows:

$$C_{P1,P2} = \frac{3+2+3+0+1}{10} = 0.9 \qquad C_{P2,P1} = \frac{0+0+3+1+0}{10} = 0.4$$

$$C_{P1,P3} = \frac{3+2+3+0+1}{10} = 0.9 \qquad C_{P2,P3} = \frac{3+0+3+1+1}{10} = 0.8$$

$$C_{P1,P4} = \frac{0+2+0+1+1}{10} = 0.4 \qquad C_{P2,P4} = \frac{0+2+0+1+1}{10} = 0.4$$

$$C_{P1,P5} = \frac{0+2+0+1+1}{10} = 0.4 \qquad C_{P2,P5} = \frac{0+0+0+1+0}{10} = 0.1$$

$$C_{P1,P6} = \frac{0+2+0+0+1}{10} = 0.3 \qquad C_{P2,P6} = \frac{0+0+0+1+0}{10} = 0.1$$

Figure 3-6 - Sample Calculation of Concordance Indices

The Concordance Matrix is thus:

	P1	**P2**	**P3**	**P4**	**P5**	**P6**
P1	X	0.9	0.9	0.4	0.4	0.3
P2	0.4	X	0.8	0.4	0.1	0.1
P3	0.1	0.6	X	0.3	0.3	0.3
P4	0.7	0.9	0.7	X	0.5	0.4
P5	0.7	0.9	0.9	1.0	X	0.6
P6	0.7	0.9	0.9	1.0	1.0	X

Figure 3-7 - Concordance Matrix

Sample calculation for some of the Discordance Indices are shown in Figure 3-8. Note that $\delta = 20$ (Maximum difference on any one criterion)

$$D_{P1,P2} = \frac{6}{20} = 0.3 \qquad D_{P2,P1} = \frac{15}{20} = 0.75$$

$$D_{P1,P3} = \frac{6}{20} = 0.3 \qquad D_{P2,P3} = \frac{5}{20} = 0.25$$

$$D_{P1,P4} = \frac{10}{20} = 0.3 \qquad D_{P2,P4} = \frac{20}{20} = 1.0$$

$$D_{P1,P5} = \frac{10}{20} = 0.5 \qquad D_{P2,P5} = \frac{20}{20} = 1.0$$

$$D_{P1,P6} = \frac{15}{20} = 0.75 \qquad D_{P2,P6} = \frac{20}{20} = 1.0$$

Figure 3-8 - Calculation of Discordance Indices

The Discordance Matrix is thus:

	P1	**P2**	**P3**	**P4**	**P5**	**P6**
P1	X	0.3	0.3	0.5	0.5	0.75
P2	0.75	X	0.25	1.0	1.0	1.0
P3	0.5	0.25	X	1.0	1.0	1.0
P4	0.75	0.3	0.3	X	0.25	0.5
P5	0.5	0.3	0.3	0.0	X	0.25
P6	0.5	0.15	0.15	0.0	0.0	X

Figure 3-9 - Discordance Indices

Solution

The aim of ELECTRE 1 is to isolate the kernel $G(\hat{c}, \hat{d})$ of the outranking graph and this depends on the concordance and discordance thresholds chosen.

If $\hat{c} = 1.0$, and $\hat{d} = 0.0$, the kernel of the graph consists of P1, P2, P3, and P6. The only options outside the core are P4 and P5 (both are outranked by P6), i.e.
For G(1.0,0) : P6 > P4, P6 > P5, P5 > P4,
Note that, for ELECTRE I, P1, P2, and P3 are incomparable between themselves and all others

If $\hat{c}$ is reduced to 0.9, and $\hat{d}$ is increased to 0.15, the kernel reduces to two options - P1 and P6, since now P6 outranks P2 and P3, i.e.
For G(0.9,0.15) : P6 > P2, P6 > P3, P6 > P4, P6 > P5
Note that here P1 remains incomparable with the other options

Only when the thresholds reach $\hat{c}=0.7$ and $\hat{d}=0.5$ does one option (P6) outrank all others and the kernel of the graph is reduced to one element, i.e.

For G(0.7,0.5): P6 > All Others

However, normally the process would stop before the thresholds reached these values as the outranking relationship loses much of its realism once the concordance index goes below 0.8 and the discordance goes above 0.3

3.5 ELECTRE II

General

The ELECTRE II system (Roy and Bertier, 1971, 1973) differs from ELECTRE I in that it ranks the options from best to worst, rather than simply establishing an initial kernel. It gives a complete ordering of non-dominated options, unlike ELECTRE I which provides a partial ordering of the non-dominated set.

Here, as with ELECTRE I, option *a* is preferred to option *b* if and only if both the concordance and discordance conditions are satisfied. However, unlike ELECTRE I, this model utilises two separate outranking relationships - a strong outranking relationship S^F, and a weak outranking relationship S^f each defined by different thresholds.

Establishing Strong and Weak Outranking Relationships

For any given pair of options (a,b), ELECTRE II establishes an outranking of a over b if the twin conditions of concordance and non-discordance are met, that is, the concordance measure c(a,b) is above a certain minimum level of acceptability and the discordance measure d(a,b) is below a set level of allowable discordance.

The method is based on the use of two different outranking relationships:

- a *strong outranking relation*, S^F, gives a very firm basis to the assertion that option *a* outranks option *b* (i.e. a strong consensus)
- a *weak outranking relation*, S^f, gives a less firm basis to the assertion that option *a* outranks option *b* (i.e. a relatively low level of consensus)

In order to define S^F and S^f, let c^+, c^0 and c^- represent three decreasing levels of concordance such that $0 \le c^- \le c^0 \le c^+ \le 1$. Also, let d_1 and d_2 represent two decreasing levels of discordance such that $0 \le d_2 \le d_1 \le 1$.

The conditions for strong and weak outranking are then defined as follows:

- Strong Outranking: aS^Fb requires

$$c(a,b) \geq c^+ \tag{3-7}$$

and

$$g_j(a) - g_j(b) \leq d_1 \; \forall \, j \tag{3-8}$$

and

$$\frac{P^+(a,b)}{P^-(a,b)} \geq 1 \tag{3-9}$$

where

P^+ represents the sum of the weights of the criteria on which option a is preferred to option b, and P^- represents the sum of the weights of the criteria on which option b is preferred to option a.

and / or

$$c(a,b) \geq c^0 \tag{3-10}$$

and

$$g_j(a) - g_j(b) \leq d_2 \; \forall \, j \tag{3-11}$$

and

$$\frac{P^+(a,b)}{P^-(a,b)} \geq 1 \tag{3-12}$$

- Weak Outranking: aS^fb requires

$$c(a,b) \geq c^- \tag{3-13}$$

and

$$g_j(a) - g_j(b) \leq d_1 \; \forall \, j \tag{3-14}$$

and

$$\frac{P^+(a,b)}{P^-(a,b)} \geq 1 \tag{3-15}$$

From these two relationships, the graphs for both the strong and weak outranking relationships can be constructed. These graphs then form the basis for the following iterative procedure which determines the ranking of all options.

The Ranking procedure

The ELECTRE II model produces a ranking of all options from first to last. It first establishes three pre-orders, two of which are total pre-orders V_1 and V_2, and a partial pre-order $\overline{V}$.

Construction of the First Total Pre-order, V_1

Before the ranking operation can proceed, all possible circuits within the strong outranking graph G_F must be eliminated. Since the options forming a circuit are of equal ranking, where they exist, they should be substituted by one equivalent option, thereby reducing the effective number of options considered. The reduced outranking graph containing no circuits is denoted $G_{F'}$.

The first complete pre-order V_1, is obtained using the 'direct ranking' procedure as follows:

- This algorithm is an iterative process. At each step l (initially l=0), the options already ranked are removed from the graph $G_{F'}$ representing the strong outranking relationships between the remaining options. For these, the strong outranking relationships between them are provided by the graph Y_l, which is a sub-graph of $G_{F'}$, the graph giving the strong outranking relationships for all options, having eliminated all possible circuits.
- A_l represents the best options within step l which will receive the ranking l+1
- Within the graph Y_l, all options not strongly outranked by any other option form the set *D*.
- Those options within *D* that are linked to each other by a weak outranking relationship constitute the set *U*.
- The set *B* consists of all options from *U* not weakly outranked by any other from within *U*.
- For a given step l, The single option or group of equally ranked options is defined by the union of the sets (*D* - *U*) and *B*. The set *D* - *U* represents all the options which:

(i) have not yet been ranked,
(ii) are not outranked by any option within Y_l, which consists of those options from the reduced strong outranking graph $G_{F'}$ that have not as yet been graded, and
(iii) are not connected by a weak outranking relationship.

- The set *B* represents all options which satisfy conditions (i) and (ii) but which, in addition must also

(iv) weakly outrank the other options that fulfil conditions (i) and (ii)

- All options ranked within the lth step are assigned the rank $l+1$. In this way, for each option, a corresponding rank is obtained by the direct ranking procedure V_1, and if the rank of option a is smaller than the rank of option b, i.e. if $r_1(a) < r_1(b)$, this indicates that option a is better than option b.

The options ranked at the lth step are removed from the strong outranking graph, leaving the sub-graph of options Y_{l+1} for consideration within the next step, i.e.

$$Y_{l+1} = Y_l - A_l$$

- Finally, if, in continuing to the next step $l+1$, it is found that Y_{l+1} is empty, the ranking procedure is terminated, otherwise the operation continues and the ranking of the remaining options determined.

Construction of the Second Total Pre-order, V_2
The second pre-order is obtained using the inverse ranking procedure. The same basic algorithm as above is utilised, with the following modifications.

- invert the direction of the strong and weak outranking graphs $G_{F'}$ and $G_{f'}$,
- the ranks obtained from the inverted graphs using the direct procedure is adjusted as follows:

'inverse procedure' rank $(r_2(a))$=1+(the number of ranks) $(r'_2(a)_{max})$ -'direct procedure' ranking $(r'_2(a))$

Construction of the Final Partial Pre-order
The intersection, in mathematical terms, of the two compete pre-orders, obtained using the direct and inverse ranking procedures, is a partial pre-order. It can allow two options to remain uncompared without affecting the validity of the overall ranking. The system for constructing the final pre-order is as follows:

- if a is preferred to b in the two complete pre-orders, then this will also be the case in the final pre-order;
- if a has an equivalent ranking to b in the one of the complete pre-orders, but is preferred to b in the other, then a precedes b in the final pre-order;
- if a is preferred to b in the first complete pre-order, but b is preferred to a in the second, then the two options are incomparable in the final pre-order.

A Worked Example of the ELECTRE II Ranking Procedure

Introduction

The following is an illustration of the ELECTRE II ranking procedure given by Maystre et al. (1994). Suppose, the strong and weak outranking relationships between 7 options have already been determined and are as follows:

	1	2	3	4	5	6	7
1	---		S^F		S^f	S^F	S^f
2	S^F	---	S^f	S^f		S^F	
3			---		S^f	S^F	S^f
4	S^F		S^F	---	S^F	S^F	S^f
5					---		
6	S^F				S^f	---	S^f
7					S^F		---

Table 3-1 - Strong and Weak Outranking Relationships

These relationships can be illustrated graphically as follows, with circuits removed:

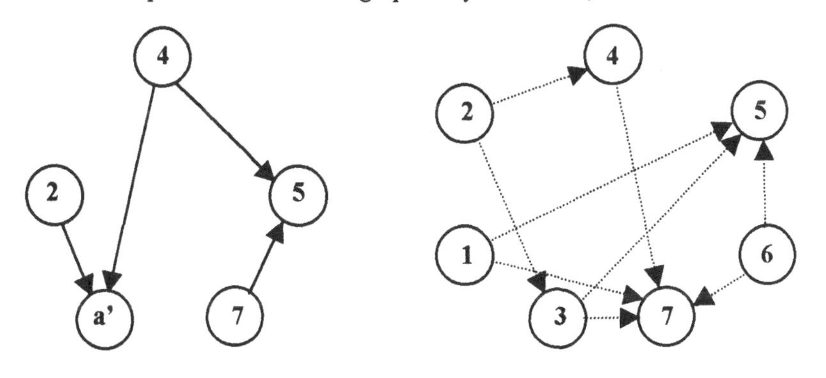

(a) Strong Outrankings (b) Weak Outrankings

Figure 3-10

- Options 1, 3 and 6 form a circuit, denoted above as a', and are taken as one option for the purposes of strong outranking. Option 2 strongly outranks Options 1 and 6. Option 4 strongly outranks Options 1, 3, 5 and 6. Option 7 outranks Option 5.

- Option 1 weakly outranks Options 5 and 7. Option 2 weakly outranks Options 3 and 4. Option 3 weakly outranks Options 5 and 7. Option 4 weakly outranks Option 7. Option 6 weakly outranks Options 5 and 7.

Direct Ranking Procedure

Step 0

($l = 0$)

Set A contains seven options, numbered 1 to 7.

$Y_l = Y_0$ = all seven options

Options 2, 4 and 7 are not strongly outranked by any of the other option, and therefore form set D.

All these three options are connected by weak outranking relationships: $2S_f4$ and $4S_f7$, therefore the same three options comprise set U.

Set D minus set U therefore contains no options.

Option 2 is not weakly outranked by any other option, and forms set B.

Therefore, the union of sets (D-U) and B contains Option **2**, which is given the rank 1 ($l + 1$), and takes no further part in the direct ranking procedure.

A_0= Option 2

Step 1

($l = 1$)

$Y_l = Y_1$ = Options 1, 3, 4, 5, 6, 7

Options 4 and 7 are not strongly outranked by any of the other option, and therefore form set D.

Options 4 and 7 are also connected by a weak outranking relationship: $4S_f7$, and comprise the set U.

Set D minus set U therefore again contains no options.

Option 4 is not weakly outranked by any of the other 5 options, and forms set B.

Therefore, the union of sets (D-U) and B contains Option **4**, which is given the rank 2 ($l + 1$), and takes no further part in the direct ranking procedure.

A_1= Option 4

step 2

($l = 2$)

$Y_l = Y_2$ = Options 1, 3, 5, 6, 7

Options 1, 3, 6 and 7 are not strongly outranked by any of the other option, and therefore form set D.

Options 1, 3, 6 and 7 are also connected by a weak outranking relationship: $1S_f7$, $3S_f7$ and $6S_f7$, and comprise the set U.

Set D minus set U therefore again contains no options.

Options 1, 3 and 6 are not weakly outranked by any of the other 2 options, and forms set B.

Therefore, the union of sets (D-U) and B contains Options 1, 3 and 6, which is given the rank 3 ($l + 1$), and takes no further part in the direct ranking procedure.

A_2= Options 1, 3 and 6

step 3
($l = 3$)
$Y_l = Y_3$ = Options 5 and 7
Option 7 is not strongly outranked by the other option, and therefore form set *D*.
Options 5 and 7 are not connected by a weak outranking relationship, therefore set *U* is empty.
Set D minus set U therefore contains Option 7.
Option 7 is not weakly outranked by Option 5, and therefore constitutes set *B*.
Therefore, the union of sets (D-U) and B contains Option 7, which is given the rank 4 ($l + 1$), and takes no further part in the direct ranking procedure.
A_3 = Option 7

step 4
($l = 4$)
$Y_l = Y_4$ = Option 5
Since only one option remains, it is ranked 5 ($l + 1$)
A_4 = Option 5

The results from the direct ranking procedure is summarised in the following table:

Step l	Y_l	D	U	B	A_l	$r_{l(l+1)}$
0	1, 2, 3, 4, 5, 6, 7	2, 4, 7	2, 4, 7	2	2	1
1	1, 3, 4, 5, 6, 7	4, 7	4, 7	4	4	2
2	1, 3, 5, 6, 7	1, 3, 6, 7	1, 3, 6, 7	1, 3, 6	1, 3, 6	3
3	5, 7	7	--	--	7	4
4	5	--	--	--	5	5

Table 3-2 - Summary of Direct Ranking Procedure

Inverse Ranking Procedure
Firstly, the outranking relationships given in Table 3-1 are inverted as follows:

	1	**2**	**3**	**4**	**5**	**6**	**7**
1	---	S^F		S^F		S^F	
2		---					
3	S^F	S^f	---	S^F			
4		S^f		---			
5	S^f		S^f	S^F	---	S^f	S^F
6	S^F	S^F	S^F	S^F		---	
7	S^f		S^f	S^f		S^f	---

Table 3-3 - Inverted Strong and Weak Outranking Relationships

These inverted relationships can be summarised as follows:

- Options 1 3, and 6 again form a circuit, and can be treated as one option for the purposes of strong outranking. Option 3 strongly outranks Option 4.. Option 5 strongly outranks Options 4 and 7. Option 6 strongly outranks Options 2 and 4.
- Option 3 weakly outranks Option 2. Option 4 weakly outranks Option 2. Option 5 weakly outranks Options 1, 3 and 6. Option 7 weakly outranks Options 1, 3, 4 and 6.

These inverted relationships are then used as before to produce the ranking $r'_{2(l+1)}$ as follows:

Step l	Y_l	D	U	B	A_l	$r'_{2(l+1)}$
0	1, 2, 3, 4, 5, 6, 7	1,3,5,6	1,3,5,6	5	5	1
1	1, 2, 3, 4, 6, 7	1,3,6,7	1,3,6,7	7	7	2
2	1, 2, 3, 4, 6	1,3,6	--	--	1,3,6	3
3	2,4	2,4	2,4	4	4	4
4	2	--	--	--	2	5

Table 3-4 - Unadjusted Inverse Rankings

The adjusted inverse rankings $r_{2(l+1)}$ are estimated using the following equation:

$$r_{2(l+1)} = 1 + r'_{max} - r'_{2(l+1)} \quad (3\text{-}16)$$

Given that r'_{max}, the number of ranking positions in the inverse ranking procedure, equals 5:

$$r_{2(l+1)} = 6 - r'_{2(l+1)} \quad (3\text{-}17)$$

The adjusted inverse rankings then be computed as follows:

A_l	$r'_{2(l+1)}$	$r_{2(l+1)}$
5	1	5
7	2	4
1,3,6	3	3
4	4	2
2	5	1

Table 3-5 - Adjusted Rankings

Since, in this case, the rankings from both procedures are identical, the final pre-order is as follows:

$2 \rightarrow 4 \rightarrow 1,3,6 \rightarrow 7 \rightarrow 5$

An Alternative Definition of Strong and Weak Outranking in ELECTRE II.

Vincke (1992) describes a different form of the model involving two concordance thresholds ($\hat{c}_1, \hat{c}_2, \hat{c}_1 > \hat{c}_2$), and two discordance thresholds ($\hat{d}_1, \hat{d}_2, \hat{d}_1 < \hat{d}_2$) in building the strong outranking relation S^F and the weak outranking relation S^f as follows :

aS^Fb ('a' strongly outranks 'b')

iff

$$c(a,b) \geq \hat{c}_1 \quad (3\text{-}18)$$

$$d(a,b) \leq \hat{d}_1 \quad (3\text{-}19)$$

and

$$\sum_{j:g_j(a)>g_j(b)} w_j > \sum_{j:g_j(a)<g_j(b)} w_j \quad (3\text{-}20)$$

aS^fb ('a' weakly outranks 'b')

iff

$$c(a,b) \geq \hat{c}_2 \quad (3\text{-}21)$$

$$d(a,b) \leq \hat{d}_2 \quad (3\text{-}22)$$

and

$$\sum_{j:g_j(a)>g_j(b)} w_j > \sum_{j:g_j(a)<g_j(b)} w_j \quad (3\text{-}23)$$

Sensitivity Testing for ELECTRE II

Those parameters of ELECTRE II that are most sensitive are:

- the concordance threshold for weak outranking ($\hat{c}_2$)
- the concordance threshold for strong outranking ($\hat{c}_1$)
- the discordance threshold for weak outranking ($\hat{d}_2$)
- the discordance threshold for strong outranking ($\hat{d}_1$)

- the criterion importance weightings (w_j)

A full case study involving the use of ELECTRE II is described in Chapter 5

3.6 ELECTRE III

General

ELECTRE III (Roy, 1978) uses the pseudo-criterion, with its indifference and preference thresholds, to explicitly make allowances for any imprecision/uncertainty in the data.

As with the first two models, ELECTRE III comprises two distinct phases:

1. constructing the outranking relation, and
2. exploiting the outranking relation.

Constructing the Outranking Relation.

ELECTRE III defines the degree of outranking of *b* by *a*, S(a,b) (or aSb) in terms of its Concordance Index C(a,b) and its Discordance Index D(a,b).

a) The following Concordance Index is computed for each ordered pair (a,b) of actions:

$$C(a,b) = \frac{1}{W} \sum_{j=1}^{n} w_j c_j(a,b) \tag{3-24}$$

where

$$W = \sum_{j=1}^{n} w_j$$

and

$$c_j(a,b) = 1 \text{ if } g_j(a) + q_j(g_j(a)) \geq g_j(b) \tag{3-25}$$

or

$$c_j(a,b) = 0 \text{ if } g_j(a) + p_j(g_j(a)) < g_j(b) \tag{3-26}$$

otherwise

$$c_j(a,b) = \frac{g_j(a) - g_j(b) + p_j(g_j(a))}{p_j(q_j(a)) - q_j(g_j(a))} \tag{3-27}$$

(note: p_j = strict preference threshold for criterion j, q_j = indifference threshold for criterion j)

C(a,b) represents the percentage of weights of the criteria that concord with the proposition '*a outranks b*'.

(note: if $q_j(g_j(a)) = p_j(g_j(a)), \forall a, j,$ then the structure becomes a semi-order one, based on the threshold model)

Thus

$$C(a,b) = \frac{1}{W} \sum_{j: g_j(a) + q_j(g_j(a)) \geq g_j(b)} w_j \tag{3-28}$$

- The definition of discordance uses a veto threshold $v_j(g_j(a))$ such that the outranking of *b* by *a* is refused if

$$g_j(b) \geq g_j(a) + v_j(g_j(a)) \tag{2-29}$$

The Discordance Index for each criterion j is as follows:

$$D_j(a,b) = 0 \text{ if } g_j(b) \leq g_j(a) + p_j(g_j(a)) \tag{3-30}$$

$$D_j(a,b) = 1 \text{ if } g_j(b) > g_j(a) + v_j(g_j(a)) \tag{3-31}$$

otherwise,

$$D_j(a,b) = \frac{g_j(b) - g_j(a) - p_j g_j(a))}{v_j(g_j(a)) - p_j(g_j(a))} \tag{3-32}$$

- The degree of credibility of outranking of *b* by *a* is defined as follows:

$$S(a,b) = C(a,b) \text{ if } D_j(a,b) \leq C(a,b), \forall j \tag{3-33}$$

otherwise

$$S(a,b) = C(a,b) \prod_{j \in J(a,b)} \frac{(1 - D_j(a,b))}{1 - C(a,b)} \tag{3-34}$$

where J(a,b) is the set of criteria for which $D_j(a,b) > C(a,b)$.

The degree of credibility of outranking is thus equal to the Concordance Index where no criterion is discordant. Where, however, discordances do exist, the Concordance Index is lowered in direct relation to the importance of those discordances.

Exploiting the Outranking Relation

The algorithm for ranking all options yields two pre-orders, each constructed in a different way. The first pre-order is obtained in a descending manner, selecting the best-rated options initially, and finishing with the worst (descending distillation).

The second pre-order is obtained in an ascending manner, selecting firstly the worst rated options, and finishing with the assignment of the best (ascending distillation).

The construction of these two pre-orders requires the qualification score for each option which are calculated using the following procedure:

Firstly, let λ_0 equal the maximum value of S(a,b) for all option pairs,

$$\lambda_0 = \max_{a,b \in A} \{S(a,b)\} \tag{3-35}$$

A cut-off level of outranking λ_1 is defined as a value close to λ_0 such that:

$$\lambda_1 = \lambda_0 - s(\lambda_0) \tag{3-36}$$

where

$s(\lambda_0)$ is called the discrimination threshold.

For a given pair of options (a,b), *a* outranks *b* at the cut-off level λ_1 if the following conditions are met:

$$aS^{\lambda_1}b \text{ iff } S(a,b) > \lambda_1 \text{ and } S(a,b) - S(b,a) > s(S(a,b)) \tag{3-37}$$

In other words, a outranks b if the degree of credibility of outranking for a over b is greater than the cut-off level, and the degree of credibility for a over b is greater than that for b over a by grater than the discrimination threshold. If these two conditions hold, it can be stated that it is significantly more credible that a outranks b than that b outranks a.

Example of Fulfillment of Outranking Requirements

Take two options 1 and 3, where:
$S(3,1) = 0.90$
$S(1.3) = 0.73$
$\lambda_0 = 0.95$
$s(\lambda) = \alpha * \lambda + \beta$
$\lambda_1 = 0.90 - s(0.95)$
$= 0.90 - 0.14$
$= 0.76$
and,
$\alpha = 0.3$
$\beta = 0.15$
therefore,
$s(\lambda) \quad = 0.3 - 0.15(0.90) = 0.165$

Option 1 outranks Option 3 because:

(i) $S(3,1) > 0.76$
(ii) $S(3,1) - S(1,3) = 0.17 > 0.165$

Thus

$3S^{\lambda 1}1$ because the degree of credibility of outranking for Option 3 over Option 1 is greater than the cut-off level λ_1, and the degree of credibility for S(3,1) surpasses the degree of credibility for S(1,3) by greater than the discrimination threshold s, thus implying that the assertion 'Option 3 outranks Option 1' is significantly more credible than the assertion 'Option 1 outranks Option 3'.

Qualification

From the outranking relationship in equation (3-37), the strength and weakness of each option a at the cut-off level λ_1 is determined as follows:

The strength of the option $p_A^{\lambda_1}(a)$ is defined by:

$$p_A^{\lambda_1}(a) = |\{b \in A / a\, S_A^{\lambda_1}\, b\}|$$

The weakness of the option $f_A^{\lambda_1}(a)$ is defined by:

$$f_A^{\lambda_1}(a) = |\{b \in A / b\, S_A^{\lambda_1}\, a\}|$$

The qualification of option *a* in relation to the set of options A, $q_A^{\lambda_1}(a)$, is defined by

$$q_A^{\lambda_1}(a) = p_A^{\lambda_1}(a) - f_A^{\lambda_1}(a)$$

This indicator expresses clearly the relative positions of the options within the set A.

Distillation

The algorithm used in the distillation proceeds on the basis of lowering the cut-off level λ from λ_0 to zero.

Two distillation procedures are employed, the downward and upward systems.

The Downward Distillation Procedure

For the first chosen cut-off level λ_1, the subset $\overline{D}_1$ of the best options within A, is obtained from

$$\overline{D}_1 = \{a \in A / q_A^{\lambda_1} = \overline{q}_A = \underset{x \in A}{\text{Max}}\ q_A^{\lambda_1}(x)\} \tag{3-38}$$

which is the subset of options within A having the greatest qualification score.

The procedure continues for those options belonging to $\overline{D}_1$, this time trying to distinguish between them on the basis of a new second outranking relation defined by the cut-off level λ_2, such that:

$$\lambda_2 = \lambda_1 - s(\lambda_1) \tag{3-39}$$

The process is repeated until the kth step is reached when the first distillate consists of only one option, called a singleton, such that:

If $|\overline{D}_1| = 1$, then one option only has been selected

If the first distillate contains more than one option, then the process continues for those options within $\overline{D}_1$, progressively lowering λ. At each step those options not having the maximum qualification score are eliminated, until the kth step is reached at which the distillate is either a singleton or two or more options are declared indistinguishable. This set, called the *first distillate*, is denoted as $\overline{C}_1$, the highest ranking option or options on the basis of the first downward distillation procedure.

If $|\overline{D}_k| > 1$ and $\lambda_k = 0$, then, on the basis of the available information, it is not possible to decide between the options remaining in $\overline{D}_k$, and each one is considered to have equal ranking for the purposes of the downward procedure.

In going from the kth step to the $(k+1)^{th}$ step, the cut-off level λ_k is replaced by λ_{k+1} using the following transformation:

$$\lambda_{k+1} = \underset{\left\{\substack{S(a,b)<\lambda_k - s(\lambda_k) \\ a,b \in D_k}\right\}}{\text{Max}} S(a,b) \tag{3-40}$$

where $s(\lambda) = \alpha * \lambda + \beta$.

The values of α and β are fixed beforehand. Values recommended by Vallee and Zielniewicz (1994) are:

$\alpha = 0.3$
$\beta = 0.15$

The second distillation then commences, using the same procedure, this time with the set of options A_1, containing all the options in A except those within $\overline{C}_1$, i.e.

$$A_1 = A / \overline{C}_1 \tag{3-41}$$

This time λ_0 is set equal to the maximum remaining degree of credibility of the outranking score S(a,b) for the remaining options. Thus $\overline{C}_2$, the singleton or group of options in the second rank according to the downward distillation procedure, is obtained. Then, the distillation procedure is applied again to $A_2 = A_1/\overline{C}_2$ to obtain $\overline{C}_3$. The algorithm proceeds until no options remain to be ranked..

This process is called a **descending distillation chain**, yielding a first complete pre-order.

The Upward Distillation Procedure

Here, for the initial derived cut-off level λ_1, the subset $\underline{D}_1$ of the worst options within A are selected such that:

$$\underline{D}_1 = \{a \in A / q_A^{\lambda_1} = \underline{q}_A = \underset{x \in A}{\text{Min}}\ q_A^{\lambda_1}(x)\} \tag{3-42}$$

which is the subset of options within A having the smallest qualification scores.

Again, if this subset contains more than one option, the procedure continues for those options belonging to $\overline{D}_1$, again trying to distinguish between them on the basis of a lower cut-off level λ_2, repeating this procedure until the set contains only one option, or the cut-off level has diminished to zero and all options remaining within the subset are declared equal. The remaining option or options are declared the first distillate of the ascending chain, and the second distillation commences on all options except those from the first distillate. Selection is again on the basis of the smallest qualification score. The upward procedure is completed when all options have been assigned a rank.

Thus, a second complete preorder called the **ascending order chain** is obtained, in which the options having the smallest qualifications are systematically set aside.

Final rankings are obtained through a combination of these two pre-orders.

The algorithm for ranking the options by the above two procedures can be described as follows:

Let A be the complete set of options to be ranked

1. Set n = 0, put or $\overline{A}_0 = A$ (descending) or $\underline{A}_0 = A$ (ascending)

2. Set $\lambda_0 = \underset{a,b \in \overline{A}_n, a \neq b}{\text{Max.}} \{S(a,b)\}$ or $\lambda_0 = \underset{a,b \in \underline{A}_n, a \neq b}{\text{Max.}} \{S(a,b)\}$

3. Put k = 0, $D_0 = \overline{A}_n$ (descending) or $D_0 = \underline{A}_n$ (ascending)

4. From among all the credibility scores that are less than $\lambda_k - s(\lambda_k)$, the one having the maximum value is chosen as follows:

$$\lambda_{k+1} = \underset{\left\{\substack{S(a,b) < \lambda_k - s(\lambda_k) \\ a,b \in D_k}\right\}}{\text{Max}} S(a,b)$$

If $\forall a,b \in D_k$, $S(a,b) > \lambda_k - s(\lambda_{k+1})$, put $\lambda_{k+1} = 0$.

5. The λ_{k+1}-qualification scores for all options within D_k are calculated

6. The maximum or minimum λ_{k+1}-qualification score is obtained: $\overline{q}_{D_k}$ (descending) or $\underline{q}_{D_k}$ (ascending)

7. The following set is then obtained

$$\overline{D}_{k+1} = \{a \in D_k / q_{D_k}^{\lambda_{k+1}}(a) = \overline{q}_{D_k}\} \text{ (descending)}$$

or

$$\underline{D}_{k+1} = \{a \in D_k / q_{D_k}^{\lambda_{k+1}}(a) = \underline{q}_{D_k}\} \text{ (ascending)}$$

8. If $|\overline{D}_{k+1}| = 1$ or $|\underline{D}_{k+1}| = 1$ or $\lambda_{k+1} = 0$, proceed to step 9, otherwise put k = k+1, $D_k = \overline{D}_k$ (descending) or $D_k = \underline{D}_k$ (ascending) and go to step 2.

9. $\overline{C}_{n+1} = \overline{D}_{k+1}$ is the group of options carried through the $(n+1)^{th}$ downward distillation, termed the $(n+1)^{th}$ distillate of the downward procedure.

$\underline{C}_{n+1} = \underline{D}_{k+1}$ is the group of options carried through the $(n+1)^{th}$ upward distillation, termed the $(n+1)^{th}$ distillate of the upward procedure.

Put $\overline{A}_{n+1} = \overline{A}_n / \overline{C}_{n+1}$ (descending) or $\underline{A}_{n+1} = \underline{A}_n / \underline{C}_{n+1}$ (ascending)
If $\overline{A}_{n+1} \neq \emptyset$ or $\underline{A}_{n+1} \neq \emptyset$, put n = n+1 and proceed to step 2
Otherwise, end the distillation.

A full case study involving the use of ELECTRE III is described below in Chapter 6

Sensitivity Testing for ELECTRE III

Those parameters of ELECTRE III which are most sensitive to a robustness analysis are as follows:

- the indifference thresholds for each criterion (q_j)
- the preference thresholds for each criterion (p_j)
- the veto thresholds for each criterion (v_j)
- the criterion importance weightings (w_j)
- the discrimination threshold ($s(\lambda)$)

3.7 ELECTRE IV

General

ELECTRE IV (Roy and Hugonnard, 1982), as with ELECTRE III, is based a family of pseudo-criteria. It's objective is to rank the options, but without introducing any criterion importance weightings. The model avoids weights by assuming that no preference structure should be based on the greater or lesser importance of the criteria. No single criterion may dominate the decision-making process.

Methodology for the Pairwise Comparison of Options within ELECTRE IV

ELECTRE IV utilises five parameters (S_q, S_C , S_p , S_S and S_V) to construct fuzzy outranking relationships. In order that the pairwise comparison of options *a* and *b* for each of the relevant criteria can be aggregated to form an overall outranking relation, the following notation is introduced:

- $m_p(b,a)$: the number of criteria for which option *b* is strictly preferred to *a*.
- $m_q(b,a)$: the number of criteria for which option *b* is weakly preferred to *a*
- $m_i(b,a)$: the number of criteria for which option *b* is judged indifferent to *a*, even though its measured performance might be better.
- $m_o(b,a) = m_o(a,b)$: the number of criteria on which options *a* and *b* perform identically.

It is clear that the following relationship must hold for all pairs (a,b), with m being the total number of criteria:

$$m = m_p(b,a) + m_q(b,a) + m_i(b,a) + m_o(b,a) + m_i(a,b) + m_q(a,b) + m_p(a,b)$$

Every pair of options is assigned to one and only one of the above seven cases.

The outranking relationships proposed by Vallee and Zielniewicz (1994) are defined in terms of the above seven computations as follows:

1. Quasi-dominance S_q: *b* outranks *a* with ***quasi-dominance*** if and only if:
- there exists no criterion on which *a* is either weakly or strongly preferred to *b*, and
- the number of criteria for which *a* is considered indifferent to *b* must be less than the number of criteria where *b* is either indifferent, weakly or strongly preferred to *a*.

$$b \; S_q \; a \Leftrightarrow m_p(a,b) + m_q(a,b) = 0 \text{ and } m_i(a,b) < m_i(b,a) + m_q(b,a) + m_p(b,a)$$

Pairwise score differences unfavourable to *b*, but below the threshold of indifference, will, if they are numerous enough, prohibit the outranking of *a* by *b*. It takes account of what Vallee and Zielniewicz (1994) called the 'cumulative effect'.

2. Canonical dominance S_c: *b* outranks *a* with ***canonical dominance*** if and only if
- there exists no criterion on which *a* is strongly preferred to *b*,
- the number of criteria on which *a* is weakly preferred to *b* is at most equal to the number of criteria on which *b* is strictly preferred to *a*, and
- the number of criteria on which *a* has a better score than *b* must not be greater than the number of criteria on which *b* has a better score than *a*, plus one.

$$b \; S_c \; a \Leftrightarrow m_p(a,b) = 0, \text{ and } m_q(a,b) \leq m_q(b,a), \text{ and } m_q(a,b) + m_i(a,b) \leq m_i(b,a) + m_q(b,a) + m_p(b,a) + 1$$

3. Pseudo-dominance S_p: *b* outranks *a* with ***pseudo-dominance*** if and only if
- there exists no criterion on which *a* is strongly preferred to *b*, and
- the number of criteria on which *a* is weakly preferred to *b* is at most equal to the number of criteria on which *b* is weakly or strictly preferred to *a*.

$$b \; S_p \; a \Leftrightarrow m_p(a,b) = 0, \text{ and } m_q(a,b) \leq m_q(b,a) + m_p(b,a)$$

4. Sub-dominance S_s: *b* outranks *a* with ***sub-dominance*** if and only if
- there exists no criterion on which *a* is strongly preferred to *b*

$$bS_sa \Leftrightarrow m_p(a,b) = 0$$

5. Veto dominance S_v: *b* outranks *a* with ***veto-dominance*** if and only if

- there exists no criterion on which *a* is strongly preferred to *b*, and the additional condition required for pseudo-dominance is not confirmed.
- If one unique criterion exists on which *a* is strictly preferred to *b* but with the differential between the two valuations falling short of the veto threshold, veto-dominance requires that *a* is strictly preferred to *b* for the majority of criteria.

$$b\ S_v\ a \Leftrightarrow m_p(a,b) = 0$$
$$\text{but, if } m_p(a,b) = 1$$
$$b\ S_v\ a \Leftrightarrow m_p(b,a) \geq m/2, \text{ and } g_j(b) + v_j[g_j(b)] \geq g_j(a)$$

For every pair of options (a,b), the degree of credibility $S(a,b) \in [0,1]$ is calculated. It indicates to what degree the statement '*a* outranks *b*' can be confirmed. The degree of credibility is obtained by the following procedure: For all pairs of options (a,b) and on all criteria, one confirms what relationship, be it strong preference, weak preference or indifference (P_j, Q_j, I_j), links *a* to *b*. Starting with this set of relationships, one can readily calculate $m_p(a,b)$, $m_q(a,b)$, $m_i(a,b)$, $m_o(a,b)$, $m_q(b,a)$ and $m_p(b,a)$ for the purposes of deducing what relationships from among S_q, S_c, S_p, S_s, S_v links *a* and *b*. If more than one qualifies, the strongest dominance relationship is selected.

For every dominance relationship, the degree of credibility of outranking associated with it has been identified by Vallee and Zielniewicz (1994) using the following subjectvely based estimates:

- if aS_qb, then $S(a,b) = 1$
- if aS_cb, then $S(a,b) = 0.8$
- if aS_pb, then $S(a,b) = 0.6$
- if aS_sb, then $S(a,b) = 0.4$
- if aS_vb, then $S(a,b) = 0.2$

If none of the above five relationships link a and b, then $S(a,b) = 0$.

By proceeding in this way for every pair of options (a,b), the full matrix of the degrees of credibility of outranking is obtained.

The process of associating a value of degree of credibility of outranking to each of the dominance relationships is only a device brought in to play to enable the application of the ranking algorithm which requires, as a starting point, a pairwise comparison matrix comprising scores within the range of zero to one.

Exploiting the Outranking Relationship

The discrimination threshold $s(\lambda)$ utilised within the ELECTRE IV method is constant and equal to 0.1, such that:

- within the first stage of classification, only the strongest dominance relationships from among those confirmed are taken into consideration
- within the second stage of classification, it is the two strongest dominance relationships that intercede in the process of ranking of the remaining options, etc.

The outranking relationship is thus exploited by means of both ascending and descending distillations, which result in two complete pre-orders, and from these come either a partial or complete final pre-order. Whether the final product is a partial pre-order (not containing a relative ranking of *all* of the options) rather than a complete pre-order depends on the level of consistency between the rankings from the two procedures.

Alternative Methodology

Building the Outranking Relation
Two relations, S_f and S_F, are built on the basis of 'common sense' considerations compatible with the lack of information on the relative importance of the criteria.

Note, $\forall j$,

1. P_j and Q_j denote the strong and weak preferences associated with the pseudo-criterion g_j, and,
2. (ii) v_j denotes the veto threshold on that criterion.(see section on pseudo-criteria)

Let us define S_F and S_f as follows:

$aS_F b$, i.e. a **strongly outranks** b
- If there exists no criterion for which *b* strongly outranks *a*, and
- If the number of criteria for which *b* is weakly preferred to *a* is at most equal to the number of criteria for which *a* either strongly or weakly outranks *b*.

$aS_f b$, i.e. a **weakly outranks** b
- If no criterion exists for which *b* is strongly preferred to *a*, but

- the second condition above in S_F is not fulfilled, i.e. the number of criteria for which *b* is weakly preferred to *a* is less than the number of criteria for which *a* is either weakly or strongly outranks *b*.

or

- If there exists a unique criterion for which *b* is strongly preferred to *a*, under condition that the following conditions hold:

a) The difference in favour of *b* is not larger than the veto threshold (or twice the strict preference threshold), and

b) *a* is strictly preferred to *b* for at least half the criteria considered.

Exploiting the Outranking Relation.
As with ELECTRE III, this is achieved by a regime of qualification, which the subsequent distillation procedure then evaluates.

The qualification procedure: Given an Outranking Relation S, defined for any set of options A, the qualification of an option *a* is the number of projects within A which are strongly outranked by *a* ('strength'), minus the number of projects/options within A which strongly outrank *a* ('weakness'), i.e.

qualification of a = (strength of a) - (weakness of a)

Thus, the initial qualification of each option can be scored.

The distillation procedure: Using, in the first instance, a downward ranking procedure, we determine, from the initial qualification scores above, the subset B_1 of actions which have the largest qualification in A for S_F.

If B_1 is a singleton (i.e. it consists of a single option), qualifications are computed ab initio from the set $A \setminus B_1$ (i.e. a new set of qualification scores are computed using the set A less B_1). From this procedure, the subset B_2 of options which have the largest qualification in $A \setminus B_1$ for S_f is determined.

Where the subsets B_1, B_2, etc. contain more than one option, the weak outranking relation S_f is utilised to separate such ex aequo cases.

Thus, in order to complete the ranking process within the subset, the qualification procedure is applied within these subsets on the basis of S_f.

As with ELECTRE III, second preorder is built by an ascending or upward ranking procedure, determining and putting to one side each time, the subsets of options which possess the smallest qualification scores.

The final ranking is derived from the combination of the two rankings obtained from the downward and upward procedures.

A full case study involving the use of ELECTRE IV is described below in Chapter 7

Sensitivity Testing for ELECTRE IV

Those parameters of ELECTRE IV xwhich are most sensitive to a robustness analysis are as follows:

- the indifference thresholds for each criterion (q_j)
- the preference thresholds for each criterion (p_j)
- the veto thresholds for each criterion (v_j)
- the discrimination threshold ($s(\lambda)$)

3.8 ELECTRE-Tri

General

ELECTRE-Tri (Yu, 1992) is a multiple criteria assignment method which assigns project options to predefined categories. The assignment of an option *a* results from a comparison of *a* with the profiles defining the limits of the categories.

Assume F denotes the set of indices of the criteria $g_1, g_2, \ldots, g_m$, ($F = \{1, 2, \ldots, m\}$), and B the set of indices of the profiles defining p+1 categories, ($B = \{1, 2, \ldots, p\}$), b_h being the upper limit of category C_h and the lower limit of the category C_{h+1}, h = 1, 2, …, p. It is assumed that criteria are monotonically increasing, with preference increasing with increasing criterion value.

Building the outranking relation

ELECTRE-Tri builds an outranking relation S which confirms or rejects the assertion aSb_h which implies that '*a* is at least as good as the reference option b_h'. As with ELECTRE III and IV, preferences are defined through pseudo-criteria. The indifference and preference thresholds ($q_j(b_h)$ and $p_j(b_h)$) constitute the intra-criterion preferential information, and reflect the imprecise nature of the valuations $g_j(a)$. $q_j(b_h)$ specifies the largest difference $g_j(a) - g_j(b_h)$ that preserves indifference between *a* and b_h on the criterion g_j; $p_j(b_h)$ represents the smallest difference $g_j(a) - g_j(b_h)$ compatible with a preference in favour of *a* on criterion g_j.

In order to confirm the statement aSb_h, two conditions must be complied with:

- concordance: for the outranking of b_h by *a* to be accepted, a sufficient majority of criteria should be in favour of this assertion,
- non-discordance: when the concordance condition holds, none of the criteria in the minority should oppose the assertion aSb_h in too strong a manner.

Two following two inter-criterion parameters are utilised in the construction of the outranking relation *S*:

- the set of criterion weightings ($k_1, k_2, \ldots, k_m$) is used as part of the calculation of concordance through computation of the relative importance of the coalition of criteria supporting the assertion that *a* outranks b_h.
- the set of veto thresholds ($v_1(b_h)$, $v_2(b_h)$, $v_m(b_h)$) is used in the discordance test, representing the smallest difference which will veto or counteract the outranking of *a* by b_h.

ELECTRE-Tri builds an index $\sigma(a,b_h) \in [0,1]$ that represents the degree of credibility of the assertion that *a* outranks b_h, $\forall a \in A$, $\forall h \in B$. The assertion aSb_h is considered to be valid if $\sigma(a,b_h) \geq \lambda$, where λ is a 'cut-off threshold', such that $\lambda \in [0.5,1]$.

$\sigma(a,b_h)$ is estimated as follows:

a) compute the partial concordance index $c_j(a,b_h)$, $\forall j \in F$:

$$c_j(a,b_h) = 0, \text{ if } g_j(b_h) - g_j(a) \geq p_j(b_h)$$
$$c_j(a,b_h) = 1, \text{ if } g_j(b_h) - g_j(a) \leq q_j(b_h)$$

otherwise

$$c_j(a,b_h) = (p_j(b_h) + g_j(a) - g_j(b_h)) / (p_j(b_h) - q_j(b_h)) \tag{3-43}$$

b) compute the overall concordance index

$$c(a,b_h) = \sum_{j \in F} k_j c_j(a,b_h) / \sum_{j \in F} k_j \tag{3-44}$$

c) compute the discordance indices

$$d_j(a,b_h) = 0, \text{ if } g_j(a) \leq g_j(b_h) + p_j(b_h)$$
$$d_j(a,b_h) = 1, \text{ if } g_j(a) > g_j(b_h) + v_j(b_h)$$

otherwise

$$d_j(a,b_h) \in [0,1] \tag{3-45}$$

d) compute the credibility index

$$\sigma(a,b_h) = c(a,b_h) \prod_{j \in \bar{F}} \frac{1 - d_j(a,b_h)}{1 - c(a,b_h)} \tag{3-46}$$

where

$$\bar{F} = \{ j \in F : d_j(a,b_h) > c(a,b_h) \}$$

Exploiting the outranking relation

The values of $\sigma(a,b_h)$, $\sigma(b_h,a)$ and λ determine the preference situation between *a* and b_h:

- $\sigma(a,b_h) \geq \lambda$ and $\sigma(b_h,a) \geq \lambda \Rightarrow aSb_h$ and $b_hSa \Rightarrow aIb_h$, i.e. a is indifferent to b_h,
- $\sigma(a,b_h) \geq \lambda$ and $\sigma(b_h,a) < \lambda \Rightarrow aSb_h$ and not $b_hSa \Rightarrow a > b_h$, i.e. a is preferred to b_h (weakly or strongly)
- $\sigma(a,b_h) < \lambda$ and $\sigma(b_h,a) \geq \lambda \Rightarrow$ not aSb_h and $b_hSa \Rightarrow b_h > a$, i.e. b_h is preferred to a (weakly or strongly)
- $\sigma(a,b_h) < \lambda$ and $\sigma(b_h,a) < \lambda \Rightarrow$ not aSb_h and not $b_hSa \Rightarrow aRb_h$, i.e. a is incomparable to b_h.

Two assignment procedures are then available

Pessimistic procedure

a) compare a successively to b_i, for $i = p, p\text{-}1, \ldots, 0$,

b) b_h being the first profile such that aSb_h,

c) assign a to category C_{h+1} $(a \rightarrow C_{h+1})$

The direction of the ranking obtained from the pessimistic procedure is from best to worst.

Optimistic procedure

d) compare a successively to b_i, for $i = 1, 2, \ldots., p$,

e) b_h being the first profile such that $b_{h\backslash} > a$,

f) assign a to category C_h $(a \rightarrow C_h)$

In this case, the direction of the ranking obtained goes from worst to best.

In order to summarise the results from the two procedures, a table can be constructed in which the options are referred to in terms of the categories to which they are assigned by the two procedures.

Whatever assignment procedure is utilised, the following seven requirements must be met:

- No option can be indifferent to more than one reference option.
- Each option must be assigned to one reference category only (*uniqueness / unicity*).
- The assignment of any one option to its allotted category is not dependent on the assignment of any of the other options (*independence*).
- The procedure for assigning options to categories must be entirely consistent with the design of the reference options themselves (*conformity*).
- When two options have the same outranking relationship with a given reference option, they must be assigned to the same category (*homogeneity*).

- It option a' outranks a, then a' must be assigned to a category at least as good as the one to which a is assigned (*monotonicity*).
- The grouping together of two neighbouring categories must not cause the alteration of options to categories not affected by the alteration (*stability*).

ELECTRE Tri is thus not a direct pairwise methodology. For each option, the outranking relationships derived are with categories not with the other options under consideration. It thus tends to be less sensitive than pairwise based ELECTRE methods to the presence of 'clones', i.e. options lying very close to each other on their criterion valuations.

3.9 ELECTRE 1S

General

The ELECTRE IS method (Roy and Skalka, 1985) is an adaptation of ELECTRE I to fuzzy logic, permitting the use of pseudo-criteria within it. As with ELECTRE I, selecting the best option requires that the complete set of options A be partitioned into two subsets, the first, the kernel N, which comprises those options not outranked by any other, and the second, A/N, containing all those options outranked by at least one of the other options. The best option is chosen from those contained within the kernel.

The concordance indices

The concordance indices for each criterion are defined for a given pair of options (a_i, a_k) in similar fashion to ELECTRE III as follows:

$$c_j(a_i,a_k) = 0 \Leftrightarrow p_j < g_j(a_k) - g_j(a_i) \tag{3-47}$$
$$0 < c_j(a_i,a_k) < 1 \Leftrightarrow q_j < g_j(a_k) - g_j(a_i) \leq p_j \text{ (by linear interpolation)} \tag{3-48}$$
$$c_j(a_i,a_k) = 1 \Leftrightarrow g_j(a_k) - g_j(a_i) \leq q_j \tag{3-49}$$

The overall concordance index is defined as follows:

$$C_{ik} = \frac{\sum_{i=1}^{m} w_j c_j(a_i,a_k)}{\sum_{j=1}^{m} w_j} \tag{3-50}$$

The values of the overall concordance indices derived are identical tothose obtained from ELECTRE III.

The discordance indices

The discordance indices for each criterion are binary in nature, in contrast to ELECTRE III, where the values derived can vary from zero to one. For each criterion, the indices are derived as follows:

$$d_j(a_i,a_k) = 0, \text{ if } g_j(a_k) - g_j(a_i) < v_j\,(a_i,a_k) - q_j\,(a_i,a_k)\,.\,\frac{1-C(a_i,a_k)}{1-c} \tag{3-51}$$

otherwise

$$d_j(a_i,a_k) = 1 \tag{3-52}$$

where c is the concordance threshold.

The overall concordance indices, which has a structure unlike that for any other of the ELECTRE models detailed above, is in the following binary form (0,1):

$$D_j(a_i,a_k) = 0, \textit{ if } d_j(a_i,a_k) = 0\ \forall j = 1,\ldots,m \tag{3-53}$$

(m = number of criteria)
otherwise

$$D_j(a_i,a_k) = 1 \tag{3-54}$$

Establishing an outranking relationship

The outranking relationship is binary in nature, and is calculated is follows:

$$S(a_i,a_k) = 1, \text{ if } C\,(a_i,a_k) \geq c \text{ and } D\,(a_i,a_k) = 0 \tag{3-55}$$

otherwise

$$S(a_i,a_k) = 0 \tag{3-56}$$

This outranking relation, by conserving overall indices of both concordance and discordance, mirrors the approach of ELECTRE I. It makes no reference to an overall credibility index, of the type used in ELECTRE III, which combines the two indices to obtain a single value.

The membership of the kernel depends on satisfying the following two conditions:

- Any option within the kernel cannot be outranked by any other option within the kernel (internal stability).
- Any option not within the kernel must be outranked by at least one of the options within the kernel (external stability).

If more than one option exists within the kernel, again, as with ELECTRE I, the concordance threshold can be relaxed until the kernel consists of one one option - a singleton. The threshold should not, however be relaxed below the value of 0.5, in the interests of the credibility of the result obtained.

3.10 What version of ELECTRE to use?

Which general procedure should be chosen?

The three general procedures contained in the six versions of ELECTRE described thus far are:

- ELECTRE I and IS involve a selection procedure which results in the choice of a single option or group of options which are assigned to a core or kernel of preferred alternatives. They also identify those options which cannot be compared with any of the others. However, the nature of this procedure could allow a 'brilliant second' option to be excluded from the kernel because it is outranked, in this instance, by the one other option occupying the kernel. Even though this option may strongly out-perform all the others that have been excluded, the procedure will still inexorably reject it from the kernel of preferred options, though a sensitivity test may highlight this discrepancy.

- Also, if one of the options in the kernel is to be withdrawn from consideration, the entire selection procedure must be gone through again, as its exclusion may completely alter the balance of the kernel. Finally, the procedure defines less transparently than others below the boundary between the 'good' and the 'bad' options. Roy and Bouyssou (1993) termed the procedure used in ELECTRE I and IS problem type 'α'

- ELECTRE II, III and IV comprise a classification procedure which results in a ranking of all options considered in relation to each other. This procedure offers the most satisfactory overall resolution to conflicts between the options which exist at the level of the individual criteria. This is notably the case in its ability to allow options to remain incomparable with each other within the framework of the final ranking (a partial pre-order). This same characteristic also makes the procedure sensitive to the existence of 'clones', i.e. a set of options very close to each other, performing in an almost identical manner with all others. Roy and Bouyssou (1993) termed the procedure used in ELECTRE II, III and IV problem type 'γ'.

- ELECTRE Tri assigns all options to predefined categories. It allows each option to be assessed in an absolute way rather than relative to the other proposed alternatives. Because it does not involve pairwise comparison, it allows consideration of a large number of options without the consequence of a huge increase in the volume of calculations required. Used at a preliminary stage, before employing a ranking procedure, it can serve as a useful preliminary refinement or 'weeding-out' procedure in advance of the application of a ranking procedure. The selection of the reference options within the procedure is vital to the operation of the method. Whether they are imposed by the decision maker or are the result of an agreement, they are open to the possibility of rejection because of the subjective manner in which they are selected. Roy and Bouyssou (1993) termed the procedure used in ELECTRE Tri problem type 'β'.

Crisp or Fuzzy Outranking?

The development of the ELECTRE Methods, aside from the question of their classification into three distinct procedures α, β and γ, focuses on whether the outranking relationships dealt with are crisp or fuzzy. It should be recalled that:

- ELECTRE I is a relatively simple procedure, but seems, with hindsight, to miss slight differences between options, given on one hand its unique concordance and discordance thresholds, and on the other hand its binary outranking relationship (outranking, no outranking).
- ELECTRE II not only constitutes a change in procedure but equally a first step towards the greater recognition of slight differences between options through both the use of two / three concordance and two discordance thresholds, and the existence of a compound outranking relationship (no outranking, weak outranking, strong outranking).
- With ELECTRE III, IV, IS and Tri, thresholds are defined for each of the criteria separately, indicating the boundaries of indifference, weak preference and strong preference. The outranking relationships derived are fuzzy, moving progressively from indifference to strong preference.

In choosing between crisp and fuzzy, two perspectives, one theoretical and one practical must be dealt with:

- The first perspective acknowledges the undeniable advantages of the incorporation of fuzziness into the ELECTRE Methods. In particular, the greater stability or robustness when faced with allowing for variations in the values of certain thresholds. With crisp values, a given change in criterion values, no matter how small, can result in the creation or destruction of an outranking relationship between two options, modifying the result noticeably; with fuzzy criteria, this modification would certainly change the indices of credibility and thus the result, but not in quite a 'brutal' manner. Roy and Bouyssou (1993) came down explicitly in favour of fuzzy outranking methods, believing that ELECTRE I and II should be referred to for historical reasons only.
- The second perspective focuses on the decision-maker's level of understanding of the tools he is employing. By understanding ELECTRE II, one can then assert that ELECTRE III is in the same vane, but more complicated. In particular, great difficulty exists in both explaining the nature of pseudo-criteria and in presenting in an understandable and graphical form all the calculations necessary to compile the degree of credibility of outranking matrix. While recognising the advantages

of fuzzy models, this perspective notes the value of a simpler and more straightforward model such as ELECTRE II in explaining the nature of outranking to the decision maker, so that he can perceive and detect better the different components of the problem being addressed.

The choice between true-criteria and pseudo-criteria must take another point into consideration. The true-criterion is utilised in the physical sciences, serving as a measure of phenomena, whereas the pseudo-criterion, by incorporating fuzziness directly into it, affirms the uncertainty associated with the evaluation of such a criterion. Within the planning of civil and environmental engineering projects, where the uncertainties inherent in criterion estimates can be significant, the choice of a fuzzy decision model is a clear recognition of the nature of the problems being confronted.

Weighting Criteria?

The assignment of weights to criteria in order to represent their relative importance is always a delicate point. Of all the ELECTRE methods, only ELECTRE IV is designed, from the start, to function without weights, on condition always that no criteria is too dominant or too insignificant. In response to those who believe the method too technocratic in short-circuiting the debate among the stakeholders regarding what relative weights the criteria should have, one can put forward the counter argument that this constitutes a simplification of the decision problem in comparison to the requirements of ELECTRE III.

Finally, it should be noted that in situations where the decision maker may feel it appropriate to apply the ELECTRE IV model, nothing prohibits the use of one of the other versions requiring information on the relative importance of the criteria, with all weightings set equal. The application of an alternative model may be useful for comparative purposes.

Some methods for obtaining criterion weightings are outlined in Chapter 4.

When Should the Appropriate ELECTRE Method be Chosen?

While it might be assumed that the choice of ELECTRE method to be used has to be made at the start of the decision process, in practice this is not the case. Selection can be deferred up to the point where the matrix of performances has been filled out. At this point, this selection is necessary, so as to permit the parameter valuations appropriate to the chosen version to be estimated.

While choosing one of the ELECTRE methods at the start of the process may have the advantage of placing the rest of the study in a clear and more transparent perspective, two arguments exist in favour of a different approach:

- At the initial stages of the study, both the decision-maker and the experts might not be particularly clear on the type of results they want to obtain. During the phase of the study when the performances of the different options are being evaluated, the experts may get a better idea of the most appropriate ELECTRE procedure to employ.

- Faced with a decision maker with relatively little interest in the methodological aspects of the process, postponing the choice avoids having to explain differences which, from a non-expert's perspective, may seem of secondary importance. Once the decision maker is reassured with the level of progress on the basic tasks such as the choice of options and the evaluation of their relative performances on a set of agreed common criteria, questions as to what version of the model is most appropriate to use will be less problematical.

What Version of ELECTRE Should You Choose?

We will not seek to identify a 'best' version of the model. What is important is that the version chosen is the most appropriate one for the format in which you wish the results to be presented, the extent to which estimates of the uncertainties associated with the criterion valuations should be incorporated into the decision, the quality of information available regarding the relative importance of the chosen criteria and the general level of sophistication of the model desired by the decision maker.

In conclusion, one should organise the choices of version into a hierarchy, starting with the nature of the procedure, then the type of criteria utilised and the nature of the outranking relationship derived, and finally select the appropriate version (see Table 3.6).

		Procedure		
Outranking	**Criteria**	α (selection)	β (allocation)	γ (ranking)
Crisp	true-criteria	I	-	II
Fuzzy	pseudo-criteria	IS	Tri	III, IV

Table 3-6

3.11 References

Brans, J.P. and Vincke, P. (1985) 'A Preference Ranking Organisation Method'. *Management Science*, 31(6), pp. 647-656.

Leclercq, J.P. (1984) 'Propositions d'extension de la notion de dominance en presence de relations d'ordre sur les pseudo-criteres: MELCHIOR'. *Revue Belge de Recherche Operationelle de Statistique et d'Informatique*, 24(1), pp. 32-46.

Maystre, L., Pictet, J., and Simos, J. (1994) *Methodes Multicriteres ELECTRE*. Presses Polytechniques et Universitaires Romandes, Lausanne.

Roy, B. (1968) 'Classement et choix en presence de points de vue multiples (la methode ELECTRE)'. *Revue Informatique et Recherche Operationnelle*. 2e Annee, No. 8, pp 57-75.

Roy, B. (1978) 'ELECTRE III: Un algorithme de classements fonde sur une representation floue des preferences en presence de criteres multiples' *Cahiers de CERO*, Vol. 20, No. 1, pp. 3-24.

Roy, B. and Bertier, P. (1971) 'La methode ELECTRE II: Une methode de classement en presence de critteres multiples'. SEMA (Metra International), Direction Scientifique, Note de Travail No. 142, Paris, 25p.

Roy, B. and Bertier, P. (1973) 'La methode ELECTRE II: Une application au media-planning'. In Ross, M. (ed.), *Operational Research 1972*, North-Holland Publishing Company, pp. 291-302.

Roy, B. and Bouyssou, D. (1993) *Aide Multicritere a la Decision: Methodes et Cas*. Economica, Paris.

Roy, B. and Hugonnard, J.-C. (1982) 'Classement des prolongements de lignes de metro en banlieue parisienne (presentation d'une methode multicritere originale)'. *Cahiers du CERO*, Vol. 24, No. 2-3-4, pp. 153-171.

Roy, B. and Skalka, J.M. (1985) 'ELECTRE IS: Aspects methodologiques et guide d'utilisation'. *Document de LAMSADE*, No. 30, Universite Paris-Dauphine, 125p, February.

Vallee, D. and Zielniewicz, P (1994) 'ELECTRE III-IV, version 3.x: Guide D'Utilisation'. *Document de LAMSADE*, No. 85, Universite Paris-Dauphine, 146p, July.

Vincke, P. (1992) *Multicriteria Decision Aid*. John Wiley.

Yu, W. (1992) 'ELECTRE Tri: Aspects methodologiques et manual d'utilisation'. *Document de LAMSADE*, No. 74, Universite Paris-Dauphine, 80p, April.

4 WEIGHTING CRITERIA FOR USE WITHIN ELECTRE

4.1 Introduction

The assignment of importance weightings to each criterion is a crucial step in the application of all versions of the ELECTRE model with the exception of ELECTRE IV. Because it is a non-compensatory decision-aid model, the interpretation of weights is different than for a compensatory system such as MAUT (Keeney and Raiffa, 1976), where they amount to being substitution rates, allowing differences in preferences, as they relate to different criteria, to be expressed on the same scale. Within ELECTRE, the 'weights' used are not constants of scale, but are simply a measure of the relative importance of the criteria involved. Vincke (1992) likens the weighting of a criterion, in this instance, to the number of votes given to a candidate in a voting procedure, with the final tally indicating the relative importance of each criterion 'candidate'.

The following four methods which can be used to weight criteria for use within ELECTRE are detailed below:

- The Direct Weighting System (Hokkanen and Salminen, 1994)
- The Mousseau System (Mousseau, 1993),
- The 'Pack of Cards' Technique (Simos, 1990), and
- The 'Resistance to Change' Grid (Rogers and Bruen, 1998)

The first two are outlined briefly. The techniques of Simos and Rogers/Bruen are explained in detail.

4.2 The Direct Weighting System

The first method is simple and straightforward; the decision maker expresses his preference by giving weights directly to the criteria. Hokkanen and Salminen (1994), in their application of ELECTRE III for choosing a solid waste system, use this method. In their study, they requested 45 decision makers to scale each criterion from 1 to 7, 7 being the most important. In parallel, they used a second procedure where each decision maker was asked to give the numeral '1' to the least important criterion, and then base the other importance values on how many times more important they appeared to be than the least important one. Hokkanen and Salminen found that, when the two sets of weights were normalised, there were only minor differences between the two procedures.

In Hokkanen and Salminen's procedure, the average of the values obtained from all decision-makers for each criterion were not always used, as, depending on the distribution of weights from the 45 individuals, the weights thus obtained might not have reflected the preferred weighting system of even one decision-maker in the group. In these situations, the final weight was obtained through majority, i.e. the final weighting value for a given criterion was the value assigned to it by the largest number of 45 decision makers.

4.3 The Mousseau System

The method devised by Mousseau (1993) is in mathematical and psychological terms, the most rigorous. It constitutes, according to Maystre et al. (1994), a real 'weighting-aid' for criteria. Put simply, it determines a range of admissible values for the importance coefficients $k_1, k_2, \ldots, k_n$ (n = number of criteria), from a set of linear inequalities on these coefficients. These inequalities are deduced from the responses of the decision makers to pairwise comparisons of a set of fictitious options, artificially contrived so that they differ from each other on, at most, three criteria.

When the inequalities are solved, a range for each criterion weighting, rather than a single value, is deduced. The boundary values of the range obtained, in the case of each criterion, can be utilised as part of a robustness analysis within ELECTRE, and the results compared.

Because of the difficulty of completing this complex method by hand, the University of Paris (Dauphine) have developed software (DIVAPIME) which will enable wider use of this technique (Mousseau, 1995).

4.4 The 'Pack of Cards' Technique

Introduction

The distinguishing feature of this weighting method lies in the linkage of a 'playing card' to each criterion. The name of each criterion is inscribed on a card, and these are then given to the decision-maker in random order. He is then asked to physically manipulate these cards in order to rank them, and to insert blank cards where appropriate in order to reinforce ranking differences where appropriate. The active participation by the decision-makers in the procedure gives them an intuitive understanding of the approach. As above with Hokkanen and Salminen's procedure, the lowest rank order is assigned the number '1', and each decision maker then proceeds upwards from this figure. The rankings are increasingly unequal as more blank cards are used.

It is important to state precisely the method for collecting the information that is required by the technique, along with the way in which Simos (1990) has proposed it should be utilised. Certain shortcomings in the method have led to a revised approach being proposed by Roy and Figueira (1998).

Collecting Information for the 'Card Game'

The 'Pack of Cards' Technique put forward by Simos can be summarised as follows:

- A number of cards are handed to the person being questioned, with the name of each criterion on a separate card, together with outline information concerning the nature of the criterion if this is deemed necessary. Thus, if n cards are handed out, there are n criteria in total being considered in the decision problem. In order to avoid influencing the decision maker, it is advisable not to assign any number to each of the individual cards. A number of blank cards are also supplied.

- The person being questioned is then asked to order the cards from 1 to n in order of increasing importance, with the criterion ranked first being the least important and the one ranked last deemed the most important. If certain criteria are, in the opinion of the decision maker, of the same importance (and therefore the same weighting), their cards are grouped together. This physical procedure results in a complete ordering of the n criteria. $\overline{n}$, the number of grades in the pre-order is noted. Most of the grades contain one criterion only. The first grade will be termed rank 1, the second rank 2, etc.

- The person being questioned is asked to consider whether the difference in importance between any two successively ranked criteria (or groups of criteria graded equally) should, on reflection, be more or less pronounced. In order that the weighting estimate will suitably reflect this greater or smaller gap in importance, they are asked to insert blank cards between two successively ranked cards (or group of cards), with the number of blanks used reflecting the size of the gap. The absence of any blank card between two successively ranked criteria corresponds to a minimum gap u being assumed to exist. One blank card corresponds to the existence to a gap equal to twice u. Two blank cards corresponds to a gap of three times u, etc.

The Method for determining the Criterion Weightings

The Simos Technique utilises the information collected by the above procedure to assign importance weights to the individual criteria. This assignment procedure, outlined previously by Maystre et al. (1994), is explained using the following example:

Consider a family F, of 12 criteria: $F = \{a, b, c, d, e, f, g, h, i, j, k, l\}$

The person being questioned assigns the cards representing the criteria to six individual groupings. Table 4.1 below illustrates the ordering of the cards that resulted, taking account of the number of blank cards between the individual groupings.

Ranking	**Grouping**	**Number of cards in each grouping**
1	$\{c, g, l\}$	3
2	$\{d\}$	1
3	*Blank Card*	1
4	$\{b, f, i, j\}$	4
5	$\{e\}$	1
6	$\{a, h\}$	2
7	$\{k\}$	1

Table 4-1

Presentation of Information provided by the 'Pack of Cards'

In order to assign a weighting score to each criterion on the basis of the information gathered, Simos suggested the following procedure:

a) the groupings are arranged from worst to best, taking the blank cards into account

b) Each criterion and each blank card is assigned a position (which Simos termed a 'weight'): the card ranked last is given position 1, the card ranked ranked second last is given position 2, etc.

c) The non-normalised weights (which Simos terms the 'average weights') are calculated for each rank by dividing the sum of the positions for that rank by the number of criteria within that rank.

d) The normalised weights (which Simos terms the 'relative weights') are calculated by dividing the non-normalised weights for a given rank by the sum of all positions of the criteria (without taking the position of the blank cards into consideration). (It should be noted in passing that the normalised weight is rounded up or down to the nearest whole number.)

Table 4.2 below illustrates the procedure for calculating the normalised weights.

Groupings	**Number of cards**	**Positions**	**Weights (Non-norm)**	**Weights (normalised)**	**Total**
$\{c, g, l\}$	3	1, 2, 3	$\frac{1+2+3}{3}=2$	$\frac{2}{86} \times 100 \approx 2$	$3 \times 2 = 6$
$\{d\}$	1	4	4	$\frac{4}{86} \times 100 \approx 5$	$1 \times 5 = 5$
Blank Card	1	(5)	---	---	---
$\{b, f, I, j\}$	4	6, 7, 8, 9	$\frac{6+7+8+9}{4}$ =.7.5	$\frac{7.5}{86} \times 100 \approx 9$	$4 \times 9 = 36$
$\{e\}$	1	10	10	$\frac{10}{86} \times 100 \approx 12$	$1 \times 12 = 12$
$\{a, h\}$	2	11, 12	$\frac{11+12}{2} = 11.5$	$\frac{11.5}{86} \times 100 \approx 13$	$2 \times 13 = 26$
$\{k\}$	1	13	13	$\frac{13}{86} \times 100 \approx 15$	$1 \times 15 = 15$
Sum	**13**	**86**			**100**

Table 4-2

Transforming the Rankings into Importance Weightings (Simos Technique)

(note that in the third column, the position score for the blank card is not accounted for in the total sum)

Consider four criteria *a*, *b*, *c* and *d*. It is assumed that, from the perspective of the person being interviewed, these are ranked in the precise order $a \rightarrow b \rightarrow c \rightarrow d$, with the gaps in importance between each adjacent pair of criteria being assumed equal. No blank cards are required. As a consequence, criterion *d* receives a weighting four times greater than *a*. These relative importance scores flow from the rather arbitrary nature of the calculation technique. It may well be that the decision-maker may feel that the ranking positions 3, 4, 5, 6 or 6, 7, 8, 9 may be more appropriate. However, the basic Simos technique does not allow discretion in relation to the choice of weighting game used.

Suppose the person being questioned wishes to progressively increase the gaps in importance scores between successively ranked criteria, To achieve this, two blanks could be placed between *a* and *b*, three blanks between *b* and *c*, and four between *c* and *d*. The resulting positions become 1, 4, 8 and 13. As a result, criterion *d* is weighted as thirteen times more important than *a*. Ranking positions of 10, 12, 15 and 19, though they may reflect the views of the person being questioned, cannot be computed using information collected by the basic Simos technique.

The procedure put forward for translating ranking positions into weights is restricted in the range of scores it can derive due to its strict definition of the relationship between the most important and least important criterion, which is set in a very deterministic way. Assuming no two criteria are graded equally at either the first or last rank, the weighting of the most important criterion will be *T* times the importance of the least important one, where *T* is the total number of cards used in the weighting game. In general, if *q* is the number of criteria in the first rank, and *p* is the number of criteria in the last, the relative importance of the two sets of criteria is equal to the following:

$$\sum_{i=0}^{q-1}(T-i))p \div (\sum_{i=0}^{p-1}(1+i))q)$$

For the example shown in Table 4.2, $T = 13$, $q = 1$ and $p = 3$. The ratio between the first and last ranked criteria is thus:

$$(13) \div ((1)\times(2)) = 6.5$$

Thus, criterion *k* is 6.5 times more important than *c*, *g* and *l*.

The decision maker questioned is not aware of this mechanism which in some cases produces unexpected results. The weighting scores used in ELECTRE are interpreted as a number of votes for each criterion. These 'votes' are added up to give the weighting for a given 'coalition' or group of people. The relationship between the maximum number of votes and the minimum number of votes given to the different criteria cannot be inferred from the pack of cards as it is presently set up. This deficiency restricts the domain of weighting games open to the decision maker.

The manipulation of equally important criteria

Returning to the example in Table 2, let us assume that the person questioned had considered that criterion *c* alone was the least important, with *g* and *l*, ranked equal, placed immediately above it. The weighting for *c* would be reduced from 2 to 1, while the weight for both *g* and *l* would increase from 2 to 2.5. Assuming other weights would remain unaltered, one might question the soundness of this increase in the weighting score for *g* and *l*. If one adopts *c* and *g* as the least important criteria, this time with *g* and *l*, ranked equal, immediately above them, a weighting score of 2.5 is obtained for the first pair, and 3.5 for the second. Here again, with all other weights unchanged, it is rather unsatisfactory that the weighting score for criterion *c* has been reduced from 4 to 3.5 simply because l, which was originally equal to c and g, has now become equal to d.

An examination of the figures in Table 4.2 offers a simple explanation for this anomaly. If one consider two criteria (or groups of criteria) ranked consecutively in the pack of cards, with no blank card between them, the weighting difference assigned to them is not constant. The weighting difference is 2 between c, g, l and d, 2.5 between b, f, i, j and e, and only 1.5 between e and a, h. This difference is, in each case, influenced, unbeknown to the person questioned, by the number of equals in the successive ranks of criteria. Moreover, this method for dealing with equally ranked criteria allows z, the ratio between the weightings of the most and least important criteria (or group of criteria), and defined by the following formula:

$$z = \sum_{i=0}^{q-1} (T-i))p \div \left(\sum_{i=0}^{p-1} (1+i))q \right)$$

to play a role which is completely outside the control of the person being questioned.

The technique for 'rounding-off' criterion weightings

The expert involved in estimating the criterion weightings runs into difficulties when the normalised scores do not add up to unity (or 100). Yet, this situation can arise with the Simos technique, as shown below in Table 4.3, with the total weights summing to a value either less than or in excess of the required unitary value.

Groupings	Number of cards	Positions	Weights (Non-norm)	Weights (normalised)	Total
$\{c, g, l\}$	3	1, 2, 3	$\frac{1+2+3}{3} = 2$	$\frac{2}{40} \times 100 \approx 5$	$3 \times 5 = 15$
$\{d\}$	1	4	4	$\frac{4}{40} \times 100 \approx 10$	$1 \times 10 = 10$
			---	---	---
Blank Card	1	(5)			
$\{b, f, I, j\}$	4	6, 7, 8, 9	$\frac{6+7+8+9}{4} = 7.5$	$\frac{7.5}{40} \times 100 \approx 19$	$4 \times 19 = 76$
Sum	**9**	**40**			**101**

Table 4-3

(The figures in Table 4.3 correspond to the first four lines in Table 4.2)

The Revised Method of Roy and Figueira

The modification of the Simos technique put forward by Roy and Figueira (Roy and Figueira, 1998) was devised for two reasons:

- to take into information concerning the relationship between the weighting of the most and least important criteria, and
- to modify the rules governing the calculation of the weights.

Introduction

The revised method gathers the same basic information as the original Simos technique, but more detail is required. The new method is readily adaptable to many different weighting situations. Decision-makers find it easier to express their weighting preferences on an ordinal rather than a numerical scale. It is a method which readily allows the weightings from more than one decision-maker to be taken into consideration.

The revised method requires additional information to be extracted from the person questioned. Within it, the decision-maker is asked how many times more important the first ranked criterion (or group of criteria) is relative to the last ranked one (or group). This value is assigned the value z. This value need not always be strictly defined. It is important, however, to show how variations in the value of z would affect the subsequent final weightings.

The algorithm which follows:

- takes into account the supplementary information forming the basis for the revision,
- eliminates the anomaly concerning the treatment of equally ranked criteria, and
- seeks to round-off their values in a precise way.

The Algorithm

The algorithm must assign weighing values to each criterion g_i, for $i = 1, \ldots., n$.
It is advisable to determine successively:

a) The *non-normalised weights* $k(1), \ldots, k(r), \ldots k(\bar{n})$ associated with each class of equally placed criteria, arranged in order of increasing importance

$\bar{n}$ = number of ranking levels
(6 in Table 4.2)

The criterion or group of criteria identified as being least important is assigned the score of 1, i.e.:

$$k(1) = 1$$

b) The *normalised weight*: k_i is designated the normalised weight of criterion g_i such that:

$$\sum_{i=1}^{n} k_i = 100$$

n = number of criteria

Calculating the non-normalised weights

Let e'_r be the number of blank cards separating class r and class r+1 of the ranking. The following parameters are then estimated:

$$e_r = e'_r + 1, \forall r = 1,\dots, \bar{n} - 1$$

$$e = \sum_{r=1}^{\bar{n}-1} e_r$$

$$u = \frac{z-1}{e}$$

(u should be calculated to 6 places of decimals)

From this, the non-normalised weight $k(r)$ is calculated for $r = 1,.., \bar{n}$ as follows:

$$k(r) = 1 + u(e_0 + \dots + e_{r-1}), \text{ with } e_0 = 0$$

(these weights should be corrected to two decimal places when rounding it up or down to its nearest value)

If there are a number of equally placed criteria on rank r, all the criteria are given the same weight $k(r)$. Table 4.4 illustrates the calculating sequence, with the data from Table 4.1, again assuming that $z = 6.5$.

Since

$$e = 1 + 2 + 1 + 1 + 1 = 6,$$

then

$$u = 5.5/6 = 0.916666$$

Rank r	Criteria of rank r	The number of blank cards following rank r e'_r	e_r	Weightings (Non-normalised) $k(r)$	Total
1	$\{c, g, l\}$	0	1	1.00	1.00 × 3 = 3.00
2	$\{d\}$	1	2	1.92	1.92 × 1 = 1.92
3	$\{b, f, I, j\}$	0	1	3.75	3.75 × 4 = 15.00
4	$\{e\}$	0	1	4.67	4.67 × 1 = 4.67
5	$\{a, h\}$	0	1	5.58	5.58 × 2 = 11.16
6	$\{k\}$	...	...	6.50	6.5 × 1 = 6.50
Sum	12	1	6	...	42.25

Table 4-4

Non-Normalised Criterion Weightings (z = 6.5)

Calculating the Normalised Weights

Let g_i be a criterion of rank r, with k'_i the non-normalised weighting score for g_i.

Put $k'_i = k(r)$

The following parameters are then estimated:

$$K' = \sum_{i=1}^{n} k_i'$$

$$k_i^* = \frac{100}{K'} k_i'$$

These values need to be corrected to 0, 1 or 3 decimal places. k_i'' is k_i' corrected to w decimal places, that is:

$w = 0$, retain no numeral after the decimal point

$w = 1$, retain one numeral after the decimal point

$w = 2$, retain two numerals after the decimal point

With this technique, one estimates the following:

$$K'' = \sum_{i=1}^{n} k_i'' \leq 100$$

$$\epsilon = 100 - K'' \leq 10^{-\omega} \times n$$

The value $\nu = 10^{\omega} \times \epsilon$ is thus an integer, at the very most equal to n.

In setting down $k'_i = k_i'' + 10^{-\omega}$ for the appropriately chosen ν criteria and k_i'' for the n-ν others, one obtains the objective of exact normalisation of the criteria in question.

In order that the rounding up/down mechanism causes a little distortion as possible to the series of weighting scores, the choice of the ν criteria which must receive an increment of $10^{-\omega}$ is made using the following procedure:

i) For each criterion g_i, calculate the two quantities:

$$d_i = \frac{10^{-\omega} - (k_i^* - k_i'')}{k_i^*}$$

$$\bar{d}_i = \frac{(k_i^* - k_i'')}{k_i^*}$$

with

$$k_i^* = \frac{100 \times k_i^{'}}{K'}$$

and

$k_i^{''}$ being the truncated estimate of k_i^*, obtained by retaining only the ωth numeral after the decimal point. d_i represents a dysfunction regarding the relative error when the weight is rounded upwards, with $\bar{d}_i$ the dysfunction when the weight is rounded downwards.

ii) Two lists L and $\bar{L}$ are assembled, defined as follows:

- The list L is made up of the set (i, d_i), arranged in order of increasing values of d_i.
- The list $\bar{L}$ is made up of the set (i, $\bar{d}_i$), arranged in order of decreasing values of $\bar{d}_i$.

Estimate the m criteria where $d_i > \bar{d}_i$, i.e. set $M = \{i / d_i > \bar{d}_i\}$, $|M| = m$.

Partition the n criteria of F into two classes F^+ *and* F^-, with the number of criteria in F^+ being equal to ν and the number of criteria in F^- being equal to $n - \nu$, i.e. $|F^+| = \nu$, and $|F^-| = n - \nu$. The criteria within F^+ will then be rounded upwards, and those within F^- rounded down. The partition of F is achieved using the following procedure:

a) If $(m + \nu) \leq n$, form F^- from the criteria of M, completing the set with the last $(n - \nu - m)$ criteria of $\bar{L}$ not belonging to M. F^+ will then be formed from the first ν criteria of $\bar{L}$ not belonging to M.

b) If $(m + \nu) > n$, form F^+ with the $(n - m)$ criteria of L not belonging to M, completing the set with the first $(\nu + m - n)$ criteria of L belonging to M. F^- will then be formed from the last $n - \nu$ criteria of L not belonging to M.

Table 4.5 shows the calculation sequence for calculating the normalised weighting of each criterion with $\omega = 1$.

Rank	Crit	Criter'n Ident. Number N	Normalised Weights k_i^* *unrounded*	Normalised Weights k_i'' **with ω = 1** *uncorrected*	Ratio d_i	Ratio $\bar{d}_i$	Norm. Weights k_i *corrected*
1	*c*	3	2.366863905	2.3	0.014000000	0.028250000	2.4 (*3*)
1	*g*	7	2.366863905	2.3	0.014000000	0.028250000	2.4 (*2*)
1	*l*	12	2.366863905	2.3	0.014000000	0.028250000	2.4 (*1*)
2	*d*	4	4.544378698	4.5	0.012239583	0.009765625	4.5
3	*b*	2	8.875739645	8.8	0.002733333	0.008533333	8.9 (*7*)
3	*f*	6	8.875739645	8.8	0.002733333	0.008533333	8.9 (*6*)
3	*I*	9	8.875739645	8.8	0.002733333	0.008533333	8.9 (*5*)
3	*j*	10	8.875739645	8.8	0.002733333	0.008533333	8.9 (*4*)
4	*e*	5	11.053254438	11.0	0.004229122	0.004817987	11.0
5	*a*	1	13.207100592	13.2	0.007034050	0.000537634	13.2
5	*h*	8	13.207100592	13.2	0.007034050	0.000537634	13.2
6	*k*	11	15.384615385	15.3	0.001000000	0.005500000	15.3
Sum	12	...	100	99.3	...	...	

Table 4-5

Determination of Normalised Weights for Each Criterion (ω = 1 and z = 6.5)

(the figures in brackets in the last column identify the ν criteria belonging to F^+ that are rounded up (ν = 7))

Let us look in more detail at the calculation sequence within Table 4.5. The information in columns 6 and 7 show that there are three criteria for which $d_i > \bar{d}_i$. These are criteria 4, 1 and 8.

Thus:

$$M \leftarrow \{4, 1, 8\} \text{ and } |M| = 3 = m.$$

since ν = 7 and n = 12

$(m + \nu) \leq n$,

therefore:

The criteria within M are members of F^-.

However, $(n - \nu - m) = 2$, therefore we must find the other 2 members of F^-. To do this we must compile the two lists L and $\bar{L}$ as follows:

Criterion Ident. Number N	Ratio d_i
11	0.001000000
10	0.002733333
9	0.002733333
6	0.002733333
2	0.002733333
5	0.004229122
8(*)	0.007034050
1(*)	0.007034050
4(*)	0.012239583
12	0.014000000
7	0.014000000
3	0.014000000

Criterion Ident. Number N	Ratio $\bar{d}_i$
12	0.028250000
7	0.028250000
3	0.028250000
4(*)	0.009765625
10	0.008533333
9	0.008533333
6	0.008533333
2	0.008533333
11	0.005500000
5	0.004817987
8(*)	0.000537634
1(*)	0.000537634

Table 4-6

Presentation of the Two Lists L and $\bar{L}$

The remaining members of F^- are the two last ranked criteria in $\bar{L}$ that do not belong to M, i.e. criteria number 5 and 11.

Therefore

$$F^- \leftarrow \{4, 1, 8, 5, 11\}$$

F^+ thus consists of the first 7 (v) criteria from within $\bar{L}$ that do not belong to M.

Therefore

$$F^+ \leftarrow \{12, 7, 3, 10, 9, 6, 2\}$$

Each of these seven criteria are rounded up by $10^{-\omega} = 0.1$, while the other criteria are rounded down. The total cumulative increase of 0.7 allows the uncorrected normalised weights which total 99.3 (see column 5 of Table 4.5) to total 100 (see column 8 of Table 4.5).

4.5 The 'Resistance to Change' Grid

Introduction

The above methods vary in complexity, and all have their drawbacks. The methods of Simos and Hokkanen / Salminen are simple and straightforward, yet lack a firm methodological basis. On the other hand, Mousseau's method, while being very complex methodologically, is limited by the quality of the contrived project alternatives 'dreamed up' for the purpose of the exercise by the decision maker (Roy, 1993). If any two alternatives differ from one another on any more than 4 criteria, the method, as it stands, becomes overly complex, the difficulties of comparison become too great, and the responses become less reliable.

Simos (1990) had initially examined a range of relatively complex weight selection techniques such as The Analytic Hierarchy Process (Saaty, 1980), Indifference Trade-Offs, (Keeney and Raiffa, 1976) Decision Analysis Weighting (Keeney and Raiffa, 1976) and The Churchman-Ackoff Method (Churchman and Ackoff, 1954), and had found very little consistency between them. He believed that this lack of consistency between different approaches made the process of criterion weighting the weak link within the decision-aid process. Simos concluded that the method chosen to elicit weights for use within ELECTRE should be simple and comprehensible to all involved in the process. He believed that a method that was easily understood would have more credibility than other more complex, less easily understood, weighting techniques.

What is thus required is a method that is relatively simple, yet with a methodological basis which insures that the weights obtained reflect the decision makers actual preferences / options in terms of what he /she conceives as important. Vincke (1992) believed that weighting methods, of the type put forward by Simos and Hokkanen et al., where weights are spontaneously awarded to the different criteria by the decision makers, did not necessarily relate directly back to the concept of criterion importance.

Rogers and Bruen (1998) put forward a theory from the area of Psychology - Personal Construct Theory (PCT) - as a way of explaining how decision makers automatically place decision criteria into a hierarchy of relative importances. The method derived within PCT to estimate the weightings is shown to be simple, comprehensible and legitimate - the three qualities called for by Simos (1990) in a weighting technique usable, in practice, within ELECTRE. The weights obtained are based on, and can be related back directly to, the concept of importance.

The general tenets of the theory are explained below, together with the significance of criterion importance within it, and how the theory can be utilised to enable the relative weightings of the criteria to be calculated.

Personal Construct Theory (PCT)

PCT was devised by George A. Kelly (Kelly, 1955). Kelly believes that the correct model of human behaviour is that of the scientist who interprets and predicts events.

He believes that everyone is a scientist trying to make sense of the world. Kelly sees the **purpose** of setting up mental structures (or 'a network of mental pathways'), within which anticipations can be made, as the **better understanding and prediction of future events** (Banister and Mair, 1968). In order to achieve this understanding of future events / proposed actions, each individual builds a mental framework of personal constructs in order to anticipate future events. Thus, events are anticipated using construct systems.

A construct is like a reference axis - a basic dimension of appraisal. It can be verbalised, as in the case of a criterion in the civil and environmental engineering field, but it can also remain unverbalised, or even unsymbolised. They are the discriminations we make - we make sense of the world by continually detecting repeated themes, categorising them, and segmenting our world in terms of them.

A person's thought processes are psychologically channeled by the ways in which they anticipate events - one checks how much sense one has made of the world by seeing how well that 'sense' enables one to anticipate it - one's personality is the way one goes about making sense of the world (Bannister and Fransella, 1986). A person anticipates events by construing their replications. These 'replications' are some form of sameness which we wish to confirm - they emerge because of our interpretation. It is Kelly's belief that detecting repeated themes underlies our making sense of the world.

Kelly noted that repetitive themes could only be appreciated in so far as a person was able to abstract similarities and differences in the continuous onrush of daily activities. In interpreting events, a person notes features of a series of events which characterise some and are particularly uncharacteristic of others - in this way he erects constructs involving similarity and contrast - if he did not, the person would have no landmarks to guide him. Thus, all constructs are, in practice, bipolar, or dichotomous in nature, in order that a construct is capable of implying either contrast and similarity in any given situation. A construct implying similarity without contrast (or vice versa) would represent a 'chaotic undifferentiated homogeneity', making the prediction of future events impossible. Put formally, Kelly states this as follows:

'A person's construction system is composed of a finite number of dichotomous constructs.' Kelly argued that it is more useful to see constructs as having two poles. It is not that they (i.e. constructs) are by definition bipolar. It is the assertion that we might find it more useful to think about them as if they were bipolar. Bipolarity in constructs allows us to logically relate the constructs between one another. Given the bipolar constructs, people will then choose for themselves that alternative in a bipolar construct through which they anticipate the greater possibility for the elaboration of their system. Thus, if people want to anticipate events , and if this is done by the personal construct system, they will move in those directions which make more sense - i.e. directions which seem to elaborate their construct systems.

Thus, Kelly's theory of constructs has so far told us that putting together a set of bipolar constructs will enable us to make predictions regarding future events. The next point is of particular importance to us. In what Kelly calls the theory's Organisation Corollary, he states that each person evolves a construction system embracing ordinal relationships between constructs. Thus, what he is implying is that construct systems are hierarchical in nature. According to Kelly, this hierarchical quality of construct systems is what makes our world a manageable place for us. The hierarchy serves a variety of purposes in science, and gives us a basis for discriminating between criteria of evaluation on the basis of importance. Kelly sees the hierarchy as pyramidal in nature, with those nearer the top of the pyramid called superordinate constructs, while those lower down termed subordinate constructs. Just as constructs are bipolar, they must also be understood as interrelated. Again, in order to avoid confusion in his predictive efforts, some organisation must be brought into the collection of constructs - Kelly believes that each person sets up a characteristically personal hierarchical system of constructs, where some are more important than others. If a person is attempting to anticipate a future event, and sets up a set of constructs to aid the appraisal, unless the constructs become organised through some structure such as this pyramid, confusion and uncertainty will affect the prediction because different subsets of constructs may yield contradictory predictions. Kelly believes that a hierarchy of constructs allows them to become interrelated and help reduce and hopefully eliminate any potential inconsistencies.

Thus, we can represent the theory in diagrammatic form as follows:

If people want to anticipate events
and
If they're using PCT
then
using constructs, they move in directions making most sense
as a result
these directions elaborate / further define their construct systems, confirming aspects of their experience, (by way of similarities and contrasts).
thus,
in order to maximise the predictive effect of constructs,
they must be ordinally related and bipolar

Construct Importance

Ten years after Kelly (1955) proposed PCT, the first major elaboration of it was attempted by Hinkle (1965). Hinkle chose to focus on the organisational qualities of constructs. He derived a method for assessing the hierarchical integration or importance of an individual's personal construct system. He contended that construct definitions must involve a statement of the location of a construct dimension in the context of a hierarchical network of construct implications - they must be put in some order according to their relative importance. Put formally, 'a construct has differential implications in a given hierarchical context'. These differential implications can result in the construct being either superordinate or subordinate, depending on its location in the hierarchy. A construct must be defined in terms of its location in a hierarchical network, that is, in terms of its subordinate and superordinate implications.

Hinkle's theory constituted a major revision of the techniques proposed by Kelly. He designed the 'Implications Grid' or Impgrid to both establish and quantify the implications for each given construct within the hierarchical network. In addition, Hinkle also developed the 'resistance to change' grid in order to demonstrate that superordinate constructs not only have more implications than subordinate constructs, but they are also more resistant to change. Both these techniques, outlined below, were used together by Hinkle to test his hypotheses, the most important of which, from the point of view of establishing a criterion weighting system on the basis of Hinkle's work, was that constructs operating at a higher level of superordination will show a correspondingly higher level of resistance to change than those at a lower hierarchical level. Let us examine the two 'grid' techniques proposed by Hinkle.

The 'Resistance -To -Change' Grid

Hinkle developed the 'Resistance -To -Change' Grid to test the hypothesis that superordinate constructs would be more resistant to change than subordinate ones, i.e., if a construct resisted a change more than the one it is being compared with, it was deemed the more important of the two. Hinkle's procedure involved asking 28 students to rank order 20 personal constructs on the basis of their resistance to change. He presented twenty constructs to each subject in all possible pairs, and for each pair the subject was asked to specify which of the two constructs he would prefer to remain the same with respect to, if he had to change in terms of only one of them. Pairwise comparison of all criteria took place within a matrix format. Using this matrix, subjects were presented with every possible pair of their constructs in turn and asked to look at each in terms of the preferred sides or poles. The subjects were asked to consider a position in which they were compelled to change to the undesired pole on one of the constructs. A tick in the matrix indicated the construct on which they would prefer to remain unmoved. Hinkle wished to demonstrate that superordinate constructs are more resistant to change than subordinate constructs. All twenty constructs were then rank ordered from highest to lowest with respect to 'resistance to change' according to the number of times each construct has been designated for no change across its nineteen pairings with other constructs, and the resistance to change would be directly related to the superordinate range of implications of the constructs, as obtained from the 'implications grid' explained below. (the superordinate range of implications of a construct comprised a simple count of the number of implications in each matrix column).

Table 4.7 below illustrates the format of the 'Resistance-To-Change' Grid.

Scoring: In a given cell within the 'resistance-to-change' matrix, the following notation was used to signify the result obtained (Fransella and Bannister, 1977):

an 'X' indicated that the column construct resists change,

a 'blank' indicated that the row construct resists change,

an 'I' indicates that independent change is not possible, and

an 'e' means that both changes would be equally undesirable.

The scoring mechanism for the matrix involves counting all the blanks on the rows and the corresponding 'X's' in the columns, since these are indications of the construct's resistance to change. For example, if, for construct '1', there are 8 row blanks, the resistance score is 8; for construct '2', there are 10 blanks in the row and one 'X' in the column, this gives a score of 11.

All scores are given at the bottom of the resistance to change matrix, and rankings of the constructs can be immediately conferred from them. Thus a 'resistance to change' score was obtained for each construct. A worked example of the 'resistance-to-change grid is given further below in Section 3.10.

Using The 'Resistance to Change' Grid To Estimate The Relative Importance Of Criteria for Civil and Environmental Engineering Projects

General

Let us look in some detail at how the relative importance of constructs relating to the physical environment can be estimated using Hinkle's 'resistance to change' grid. In the context of engineering projects, the personal constructs are denoted by the environmental, economic, social and technical criteria on which the project options are compared. Hinkle's system will allow us to estimate the relative importance of those criteria to the decision makers being questioned. The result will be a set of weights directly relateable to the relative importance of the criteria under examination.

Each of the criteria deemed relevant to the decision study by the scoping process are listed. A desirable and an undesirable side of each criterion is noted, for example:

very low levels of air pollution' / 'very high levels of air pollution

very low levels of community severance' / 'very high levels of community severance

With an environmental criterion, for example, the preferred side of each criterion is assumed to be the one minimising environmental impact (the left hand side shown in the above two cases).

Each criterion, in turn, is compared, pairwise, with each of the other ones in the project study. The decision maker is asked to look at the two given constructs and asked to indicate the side he would prefer to be on, i.e. for the relevant environmental criteria, the side minimising environmental impact - in the case of the two impacts shown above, minimum community severance and minimum air pollution impacts. He is asked if he had to change one of these impacts to its undesirable side, which one of these impacts would he be least willing to see change. The preferred impact is noted by the surveyor.

It is preferable that a positive choice be made for each pairwise comparison. However, Hinkle lists two circumstances where this may not be possible:

- Both choices appear to the decision maker to be equally undesirable.

- It is not logically possible to change one construct and, at the same time, remain the same on the other, i.e., changing one construct logically implies a change on the other in the pair.

Once all the pairings have been examined, the relative resistance to slot change of the criteria can be determined by rank ordering them in terms of the following scoring formula which takes into account the number of times each construct resisted being changed during the pairing sequence.

Scoring the Grid
For each criterion / construct, the total number of times it resisted being changed in all its pairings is noted. This is its 'resistance score'. The total number of actual choice pairings for each construct was then calculated as follows:

No. of actual choice pairings = (n-1) - (no. of logically inconsistent pairings + No. of equally undesirable pairings)
n = no. of criteria.

(A simpler though somewhat less accurate scoring method for the resistance to slot change would be to order the criteria based on the percentage of times each resisted being changed within its pairings.)

The 'Implication Grid'

General
Having constructed the original 'resistance to change' grid for the twenty personal constructs, Hinkle then constructed a 20 x 20 'implication grid' matrix, using the same criteria in an attempt to provide, within the framework of a grid, a schematic representation of the subordinate and superordinate implications that interrelate a set of constructs.

This involved presenting constructs to the subject one at a time, and asking him if he were to change on that particular construct, from one pole to another, on which of the other nineteen constructs would a change also be necessary. For example, the subject is requested to consider a given construct - say construct 1 - and asked to consider the possibility of changing from his originally preferred pole of the construct to its opposite side. The subject is then asked to identify which of the other remaining nineteen constructs would be likely to be changed by a change on this one construct alone - which of these constructs does the subject expect a change to occur as a result of knowing that they have changed poles one this one construct. Those identified by this procedure constitute the 'superordinate implications' for construct 1.

In this procedure, each construct is paired successively with every other one twice, with any one of four possible outcomes:

1). a change in Construct A implies a change in Construct B
(recorded as one 'superordinate' implication for A, and one 'subordinate' implication for B)

2). a change in Construct B implies a change in Construct A
(recorded as one 'superordinate' implication for B, and one 'subordinate' implication for A)

3). a change in either Construct implies a change in the other
(recorded as a 'reciprocal' implication for each other)

4). a change in either Construct has no implications in terms of the other, in which case nothing is recorded.

Scoring

The resulting data is entered into a symmetrical matrix to facilitate the tabulation of the number of superordinate, subordinate and reciprocal implications for each of the constructs, with the column patterns representing the superordinate implications of the various constructs. The row patterns represent the subordinate implications of the various constructs, that is, a row pattern indicates those constructs of the set which the subject could utilise to imply his polar position on a given construct, with the construct at the head of the row being the common superordinate implication of these constructs.

In a given cell within the resistance grid, as seen below in Table 4.7, the following notation was used to signify the result obtained:

1) An 'X' indicates the parallel superordinate implications of the column construct
2) An 'r' in a column indicates the reciprocal superordinate implications of the column construct,
3) A blank indicates no implication
(row entries indicate subordinate implications of row constructs)

The column for each construct was summed to indicate its first order range of implications.

Applying The 'Resistance To Change' Grid To A Set Of Decision Criteria - A Worked Example.

General

In order to illustrate how the 'resistance to change' grid operates, the authors have taken the eight economic, technical and environmental criteria used by Hokkanen and Salminen (1994) to evaluate competing solid waste management systems, and will use the grid to ordinally rank the criteria and elicit their relative weightings.

The eight criteria used by Hokkanen and Salminen were as follows:

g(1) - cost of waste treated

g(2) - technical reliability

g(3) - global warming effects

g(4) - releases with health effects

g(5) - acidificative releases

g(6) - surface water dispersed releases

g(7) - number of employees

g(8) - waste recovery rate

Let us look at these criteria in more detail.

g(1), g(2), g(7) and g(8) are economic, technical and operational criteria:

Cost of waste treated - This criterion contains all economic objectives, and is expressed in net cost per ton of waste treated.
Technical reliability - This criterion contains all technical objectives. it can only be measured qualitatively, based on expert, professional opinion.
Number of employees - This criterion is operational. It was important in Finland, as (in common with Ireland) high unemployment levels exist there.
Amount of recovered waste - This criterion is also operational, and would take account of the 'market potential' of the recovered product.

g(3), g(4), g(5) and g(6) are environmental criteria:

Global warming effects - represented the total amount of greenhouse effects in each alternative.
Releases with health effects - represented those heavy metal releases to air and water which effect health, particularly lead (Pb) and cadmium (Cd).
Acidificative releases - represented the total amount of acidificative emissions, i.e. gases causing acidification: sulphur dioxide, nitrogen oxides, hydrogen carbons, hydrogen chloride.
Surface water dispersed releases - represented the surface water releases from landfill plants. Nitrogen concentrations in leachates are high, and it is an important nutrient factor particularly for lakes. Thus, nitrogen levels are generally taken to reflect this criterion level.

Constructing the Grid
The first step involves expressing each of the above criteria as a bipolar construct:

(C1) economical waste treatment / costly waste treatment

(C2) technically reliable / technically unreliable

(C3) negligible global effects / significant global effects

(C4) negligible releases with health effects / significant releases with health effects

(C5) negligible acidificative releases / significant acidificative releases

(C6) negligible surface water dispersed releases / significant s.w. dispersed releases

(C7) high employment levels / low employment levels

(C8) high waste recovery rate / low waste recovery rate

The left hand side of each of the above 8 constructs constitute the desired pole from the perspective of the decision maker, i.e. ideally, the chosen option would treat waste cheaply, be technically reliable, generate high levels of employment, achieve a high waste recovery rate and result in negligible environmental impact under the above four headings of global and health effects and acidificative and s.w. dispersed releases, and, if one alternative satisfied all, it would be the obvious choice.

Having assembled the bipolar constructs and identified the desired side of each, the 'resistance to change' grid can now be put together as follows:

The decision maker compiling the importance weightings compares each of the 8 constructs with every other one. For each comparison, knowing the preferred pole for each of the two constructs, the decision maker is instructed that he must change from the preferred side to the unpreferred side on one, but can remain the same on the other. He completes the comparison by indicating which of the two changes is the more undesirable for him.

In the grid, an 'X' indicates that the column construct resists change; a blank indicates that the row construct resists change. There are two circumstances where it may prove impossible for the decision maker to make a choice, firstly, where the two changes appear equally undesirable to the decision maker (denoted in the grid as 'I'), and, secondly, where it is not logically possible to change poles on one construct while at the same time remaining constant on the other (denoted in the grid as 'e').

A sample 'resistance to change' grid for the above eight constructs is compiled below as follows:

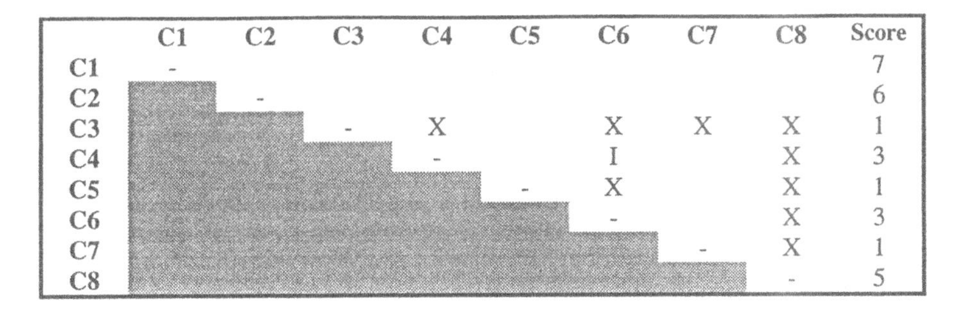

	C1	C2	C3	C4	C5	C6	C7	C8	Score
C1	-								7
C2		-							6
C3			-	X		X	X	X	1
C4				-		I		X	3
C5					-	X		X	1
C6						-		X	3
C7							-	X	1
C8								-	5

Table 4-7- Resistance to Change grid

Scoring the 'Resistance to Change' Grid

For each construct, all the blanks along the rows above the diagonal are added to all the corresponding 'X's in the columns to give the final score. Thus, for construct 1, there are 7 blank rows, so the score for resistance is 7; for construct 2, there are 6 blank rows and no 'X's, so the score is 6; construct 3 has 1 blank and no 'X's, so the score is 1; construct 4 has 2 blanks and 1 'X', so the score is 3; construct 5 has 1 blank and no 'X's, so the score is 1; construct 6 has 1 blank and 2 'X''s, so its score is 3; construct 7 has no blanks and one 'X', so the score is 1; finally, for construct 8, there are no blanks and 5 'X's, so the score is 5.

Finally, the resistance scores for each construct / criterion can be normalised to give the final weightings for each as follows:

Construct /Criterion	Description	Weights
C1	Cost of waste treated	0.26
C2	Technical reliability	0.22
C8	Waste recovery rate	0.18
C6	S.W. dispersed releases	0.11
C4	Releases with health effects	0.11
C5	Acidificative releases	0.04
C3	Global warming effects	0.04
C7	Number of employees	0.04

Table 4-8 - Final Weightings

It can be observed from the results above that three of the constructs 3, 5 and 7, violate the logical principle of transitivity, i.e. construct 3 (C3) resisted change more than construct 5 (C5), C5 resisted more than C7, and C7 resisted more than C3. Hinkle (1965) states that such instances indicate that the constructs in question are practically equal in importance. Such logic in no way undermines the credibility of the results obtained using the grid.

Analysis Of Results From The Two Grids

Hinkle found that the rank order of constructs derived from the 'resistance to change' grid was highly correlated with the rank order derived from the implications grid. Hinkle found that the higher the rank order of a given construct with respect to its measured 'resistance to change', the greater the number of superordinate implications which it carried. Hinkle interpreted his results as supporting his hypothesis that 'there is a higher level of resistance to change on superordinate constructs, because any change at this level will necessarily involve a great number of related changes'. Hinkle concludes that a superordinate construct will show greater 'resistance to change', i.e. **the more a construct resists change, the more superordinate** (i.e. higher up the hierarchy) **it is likely to be.**

Fransella and Bannister (1977) independently tested the ideas of Hinkle on the 'implications' grid and 'resistance-to-change' grid. Looking at Hinkle's most important hypothesis, that the rank order of the constructs as derived from the 'resistance to change' grid would be highly correlated with the rank order from the 'implications' grid, they examined the rank order for the 20 constructs in the original thesis in terms of their superordinate implications (column totals) and their resistance-to-change scores. The Spearman rho rank order correlation between the number of implications and their resistance to change was calculated at 0.7 - a highly significant correlation when n=20. Indeed, were it not for a very large difference between the number of implications and the resistance score for one of the constructs (construct 8 in Hinkle's original thesis), the correlation between these two measures would have been 0.75.

PCT In Engineering Decision Making

There are two potential ways of utilising PCT directly in an Civil and Environmental Engineering Decision Making system based on the Outranking Method.

The first potential use of PCT is in the elicitation of constructs / criteria. There are a number of methods used within PCT for identifying all criteria relevant to the study being undertaken. One of the most prominent of these techniques, proposed by Hinkle (1965), is a procedure called 'laddering'. This is a form of construct identification procedure which identifies their hierarchical integration. Starting with an initial preliminary list of constructs, for each one elicited, the subject is asked which pole of that construct they would prefer to be on, and why this is the case. Their answer to this question will supply a construct superordinate to the first. The subject is again asked why, and the process continues. Equally, the laddering technique can be directed downwards to more subordinate constructs by asking initially 'what constitutes' the particular construct in question, and proceeding downwards from there to more subordinate constructs.

However, at present, all criteria / constructs relevant to the civil / environmental decision process are identified using the scoping process. Scoping is defined as an early and open process for determining the significant issues related to the proposed action / project (Canter, 1995). It insures that only the more pertinent and significant criteria / factors will be included in the study, and relies upon professional judgement in the selection of economic, environmental, social and technical factors for inclusion in a project study, usually the result of expert / interdisciplinary team discussions, or of conversations with experts.

With particular reference to Environmental Impact Assessment (EIA) procedures in Ireland, the EPA have published a document which takes a wide range of project types and provides detailed guidance on the criteria that would usually be addressed when preparing an Environmental Impact Statement for the given development (EPA, 1995).

Thus, because scoping is a well established procedure for eliciting the criteria most relevant to the engineering decision process, the use of laddering or any other PCT technique for construct elicitation is not required within the present framework. However, future research in this area may point to such PCT methods as being valid, usable alternatives to scoping.

The second potential application for PCT in Engineering Decision making is, however, in the context of this author's work, much more important. Use of the ELECTRE III model requires the relative weightings of the criteria to be known. The assignment of these importance weightings to civil / environmental engineering project impacts is not, however, an area where a particular methodology has been prescribed. Most practitioners merely state that if importance weightings for the criteria are to be utilised, the rationale and methodology used should be clearly delineated (Canter, 1995). The methodologies used vary from simplistic to overly complicated. Hinkle's methodology for assessing construct importance offers some advantage over techniques at present in use within Outranking Decision models.

PCT is simpler and more generally applicable than Mousseau's weighting technique, which, though well founded in mathematical and psychological terms, is limited by the ability of the decision maker to assemble a number of contrived project alternatives in order to implement this 'weighting aid' system.

On the other hand, it is more methodologically coherent than the simpler weighting techniques of Hokkanen and Salminen (1994) and Simos (1990) referred to above. Hokkanen and Salminen's application of ELECTRE III to a Waste Management problem involved using two weighting procedures, the second of which mirrors closely that proposed by Simos. While the methods by both these authors have the advantage of being both simple and comprehensive, they do not employ a system of pairwise comparison which Saaty (1987) believes is essential for deriving criterion weightings which are measured in terms relative to one another, as is the case within ELECTRE, rather than in 'absolute terms, on a standard scale of measurement, as is the case with a 'weighted sum' decision-aid method.

Saaty (1980) believed that, where a decision maker has a significant level of expertise, as is the case with those weighting criteria within an engineering assessment, there was 'no better way' of getting those judgements down than through a systematic procedure (such as Hinkle's) which facilitated direct comparisons between criteria. Saaty (1980) found the pairwise comparison approach to have a level of consistency not available to direct methods (such as those employed by Hokkanen and Salminen (1994) and Simos (1990)) where the smallest / least important criterion is given the value 1, and all others are assigned multiples of this value. In addition, Forman (1992) noted that relative judgements, such as those made in a pairwise comparison framework are, easier to make and more meaningful than absolute judgements made within a 'direct' method.

Hinkle's method for estimating criterion importance rankings is simple and straightforward. Based on Kelly's theory of Personal Constructs, whose Organisation Corollary dictates that all criteria / constructs are hierarchically formed, Hinkle's 'resistance to change' grid allows this hierarchy to be established and the scores used as a basis for assigning relative weights. Unlike the methods of Hokkanen and Simos, weighting results obtained by Hinkle's technique can thus be validly related to the perceived relative importance levels of the criteria. It is a simple technique yet possesses a sound basis in formal psychology, directly connecting the rankings obtained to the subject's perceived relative importance weightings of the constructs / criteria.

4.6 Conclusions

The 'Resistance-To-Change' Method for estimating criterion weights (Rogers and Bruen, 1998) represents an improvement on the other methods used in non-compensatory system because:

- It is relatively simple and straightforward, and easy to use.
- It has a theoretical basis within the psychology of human preference relationships.
- The weights obtained can be directly connected, in theoretical terms, to the decision makers concept of personal importance.
- Both it and its derivatives have been used widely in Ireland within the field of market research, though in this context, where one is often dealing with large numbers of constructs, the method was found to have its limitations.
- However, given that, in the context of its use within the assessment of major civil and environmental engineering projects, only a limited number of decision makers would be required to complete the 'resistance-to-change' questionnaire, the drawbacks of the method associated with a large scale survey do not apply.
- The 'Resistance-To-Change' Method thus remains very usable within the context of all ELECTRE methods that require criterion weightings.

The method has certain operational drawbacks. If one or more criteria receives a zero score on the resistance-to-change grid, one unit must be added to all scores to enable weights to be calculated. Also, the problem of the over-complexity of the questionnaire remains if too many criteria are involved.

Where circumstances allow, it may well be the preferred course of action to combine this method with the Simos revised technique, with the grid being used to estimate the 'distances' between the criteria as an alternative to the 'blank card' technique.

4.7 References

Bannister, D. and Francella, F. (1986) *Inquiring Man: The Psychology of Personal Constructs,* Croom Helm, U.K.

Bannister, D. and Mair, J.M. (1968) *The Evaluation of Personal Constructs.* Academic Press, London.

Canter, L. (1995) *Environmental Impact Assessment,* McGraw Hill, New York.

Chuechman, C.W. and Ackhoff, R. (1954) 'An Approximate Measure of Value'. *Operations Research*, 2, 172 - 191.

Diop, O (1988) *Contribution a l'etude de la gestion des dechets solides de Dakar: analyse systemique et aide a la decision.* Thesis No.757, Ecole Polytechnique Federale de Lausanne.

Environmental Protection Agency(1995) *Advice Notes on Current Practice in the Preparation of Environmental Impact Statements.* Environmental Publications, Dublin.

Foreman, E. H. (1990) 'Multi-Criteria Decision Making and The Analytic Hierarchy Process'. *Readings in Multiple Criteria Decision Aid (ed. C Bana e Costa)*, pp295-318.Springer Verlag, Berlin.

Francella, F. and Bannister, D. (1977) *A Manual of Repertory Grid Technique.* Academic Press, London.

Hinkle, D. (1965) *The Change of Personal Constructs from the Viewpoint of a Theory of Construct Implications.* Ph.D. Dissertation, Ohio State University.

Hokkanen, J. and Salminen, S. (1994) 'Choice of a Solid Waste Management System by Using the ELECTRE III Method'. *Applying MCDA for Decision to Environmental Management.* (Ed. M. Paruccini), Kluwer Academic Publishers, Dordrecht, Holland..

Keeney, R.L. and Raiffa, H. (1976) *Decisions with Multiple Objectives: Preferences and Value Trade-Offs.*: Wiley, New York.

Kelly, G.A. (1955) *The Psychology of Personal Constructs*, Volumes 1 and 2. Norton, New York.

Maystre, L., Pictet, J., and Simos,J. (1994) *Methodes Multicriteres ELECTRE. Description, conseils pratiques et cas d'application a la gestion environmentale.* Presses Polytechniques et Universitaires Romandes, Lausanne.

Mousseau, V. (1989) *La Notion d'importance relative des criteres.* Ph.D. Dissertation, Universite Paris Dauphine

Mousseau, V. (1995) Eliciting information concerning the relative importance of criteria. *Advances in Multicriteria analysis* (Pardalos, Y., Siskos, C. and Zopounidi, C. (eds.)), pp17-43. Kluwer Academic Publishers.

Roy, B. (1993) *Methodologie Multicritere d'aide a la Decision: Methodes et Cas.* Collection Gestion, Economica, Paris.

Roy, B. and Figueira, J. (1998) 'Determination Des Poids Des Criteres Dans Les Methodes De Type ELECTRE Avec La Technique De Simos Revisee', Universite Paris-Dauphine, *Document de Lamsade 109.*

Saaty, T. L. (1980) *The Analytic Hierarchy Process.*

Mc. Graw Hill, New York.

Saaty, T. L. (1987) Rank Generation, Preservation and Reversal in the Analytic Hierarchy Decision Process. *Decision Sciencies*, Vol. 18, No. 2.

Simos, J. (1990) *Evaluer L'Impact sur L'Environment* Presses Polytechniques et Universitaires Romandes, Lausanne.

Vincke, P (1992) *Multicriteria Decision Aid.* Wiley, Chichester, U.K.

5 CASE STUDY 1
Finding the Best Location for the Galway Wastewater Treatment Plant

5.1 Introduction

Galway city is located on the west coast of Ireland. It has a population of approximately 40,000, and at present has no sewage treatment plant, with sewage being directly discharged into Galway Bay. The planned project involves the provision of a sewerage network for Galway city and its suburbs, with a main outfall in the Galway Bay area, where a treatment plant would be located. The environmental study undertaken by Consulting Engineers on behalf of Galway Corporation (McCarthy, 1992) to select the optimum location for a proposed new wastewater treatment plant in Galway City considered five potential sites within the Bay area:

- Mutton Island (MI)
- Lough Atalia 1 (shallow outfall) (LA1),
- Lough Atalia 2 (deep outfall) (LA2),
- Hare Island (HI) and
- South Park (SP).

The relative location of these sites can be seen on the map in Figure 5.1,

Mutton Island is located approximately 1km. out into the bay. Locating the treatment plant at this location would require the construction of a causeway from South Park on the southern fringe of Galway City out to the island. The trunk sewer would be on stilts over the causeway or located in a trench within it. South Park itself was also examined as a potential location. The Hare Island site is similar to the

Mutton Island proposal, as the construction of a causeway to it would also be necessary. In both cases, the causeway would be required for access during construction and afterwards for maintenance purposes. Two proposals were considered at the Lough Atalia site, one with a shallow outfall which would be located near a residential area, with the other extending the outfall into a deep water area.

Each option was assessed on the basis of the following seven mainly environmental criteria: visual impact, amenity impact, impact on birdlife, water quality impact, marine impact, construction impact and cost. Based on the information contained in the Environmental Impact Statement, the relative performance of each option on the six environmental criteria referred to above was interpreted in terms of the following five point scale:

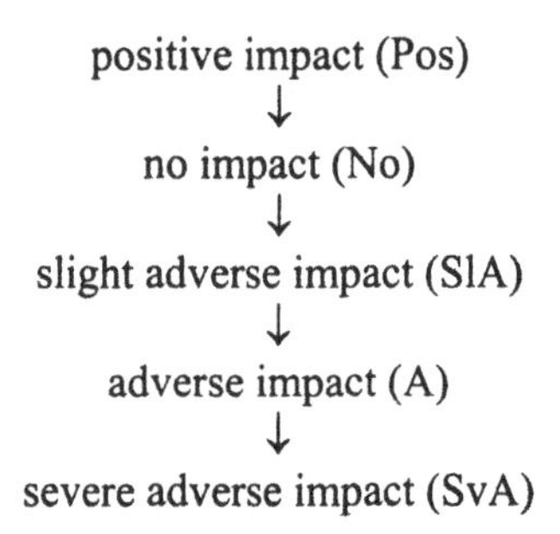

Let us define each criterion in more detail

Visual Impact (Vis)
The effect of the structure of the treatment works on the view of Galway Bay, on of the major scenic attractions on the West Coast of Ireland, is measured in this impact.

Amenity Impact (Amt)
As referred to above, some of the options require the construction of a causeway. This may add to the amenity value of the surrounding area by enabling the extension of existing walkways and by allowing a marine to be built to one side of it. Other options, however, may cause erosion of the beaches on the East Side of the Bay. This impact measures the net effect of each option on the existing amenity levels.

Impact on Birdlife (Bird)
There is a substantial population of breeding birds in the Bay area, with Mutton Island in particular designated an area of scientific interest. This impact measures the effect of the construction of the treatment works on this population

Water Quality Impact (Wat)
This impact gauges the impact of the proposed project on the bathing waters and shellfisheries in the Bay area.

Marine Impact (Mar)
The construction of a causeway would be likely to cause the disappearance of some marine flora and fauna. This impact measures the effect of each option on the marine life itself and the group depending on some species of it for their livelihood.

Construction Impact (Con)
The construction period for the project has been set by the consultants at 2.5 years. During this time, residents in the vicinity of the site will suffer some disruption. The use of explosives may also be required. This impact measures the overall constructional effect of each option on the local residents.

Cost (Cst)
This impact measures the relative cost of each of the five options.

Because of the qualitative and crisp nature of the data available from the evaluation, it was decided to use one of the simplest versions of ELECTRE available- ELECTRE II (see Section 3.5).

5.2 Criterion Valuations

The above five-point scale has been used as a basis for evaluating the five options for each of the six environmental impacts. The value given for each impact is based on the information contained in the Environmental Impact Statement commissioned by Galway Corporation (McCarthy, 1992) and a Report undertaken by Mouchel (1994) for the Save Galway Bay Group. The scores can be summarised as follows:

		Option				
		Mutton Is	**Atalia 1**	**Atalia 2**	**Hare Is**	**Sth Park**
Criterion	**Vis**	*SvA*	*SlA*	*SlA*	*SvA*	*SvA*
	Amt	*A*	*Pos*	*Pos*	*A*	*SvA*
	Bird	*SvA*	*SlA*	*SlA*	*A*	*SlA*
	Wat	*Pos*	*Pos*	*Pos*	*Pos*	*Pos*
	Mar	*SvA*	*SlA*	*SlA*	*SvA*	*N*
	Con	*SvA*	*SlA*	*A*	*SvA*	*SvA*

Table 5-1 - Environmental Criterion Grades

The cost of each option was evaluated in relative terms as follows:

	Mutton Is	**Atalia 1**	**Atalia 2**	**Hare Is**	**Sth Park**
Cst	*Low cost*	*high cost*	*higher cost*	*highest cost*	*lowest cost*

Table 5-2 - Relative Cost of Options

5.3 Weighting System

The assignment of weights to the various criteria was undertaken using a group of experts in the area of civil engineering design. Using a direct weighting technique (Hokkanen and Salminen, 1997) each respondent scored each criterion on a scale of 1 to 8, with the most important being assigned 8. Also as with Hokkanen and Salminen (1997), the first and third quartiles were taken as the minimum and maximum weights respectively, with the median used to give us the most likely valuations. These values have been estimated as follows:

	Median weighting	Lower Quartile weighting	Upper Quartile weighting
Vis	2	1.75	4
Amt	3	1	5
Bird	4	2	5.75
Wat	8	6.75	8
Mar	6	5	6
Con	1	1	2.5
Cst	1.5	1	7.25

Table 5-3 - Criterion Importance Scores

These scores were used as a basis for the following three weighting systems:

Weighting System No.1
The median scores are used for all criteria, indicating the most likely weighting for each effect.

Weighting System No. 2
Water Quality and Marine Impacts, the two highest scoring environmental effects in the median set, are scored at their lower quartile values, while the other four environmental impacts are set at their upper quartile values. Cost is left at its median value. This system gauges the effect of varying the relative weights of the environmental criteria on the final option choice.

Weighting System No. 3
Cost is placed at its upper quartile value and all the environmental impacts are set at their lower quartile values. This system gauges the effect on option choice of greater emphasis on the cost of the project relative to its environmental impact.

5.4 Concordance and Discordance Matrices

Weighting System No. .1

Introduction

Using the median scores for each criterion weight shown in Table 5.3, the following normalised weights were calculated:

Criterion	Vis	Amt	Bird	Wat	Mar	Con	Cst
Median Weighting	0.078	0.118	0.157	0.314	0.235	0.039	0.059

Table 5-4 - System No. 1 Weighting

The resulting Concordance and Discordance matrices are as follows:

	MI	LA1	LA2	HI	SP
MI	-	0.373	0.373	0.843	0.549
LA1	0.941	-	1	1	0.706
LA2	0.941	0.902	-	1	0.706
HI	0.941	0.314	0.314	-	0.549
SP	0.882	0.53	0.53	0.882	-

Table 5-5 - Concordance matrix for Weighting System 1

	MI	LA1	LA2	HI	SP
MI	-	0.75	0.75	0	0.75
LA1	0.25	-	0	0	0.5
LA2	0.5	0.25	-	0	0.75
HI	0.75	0.75	0.75	-	1
SP	0.25	1	1	0.25	-

Table 5-6 - General Discordance matrix

The Concordance and Discordance thresholds which determine the strong and weak outranking relationships were chosen as follows:

$c^+ = 0.85$
$c^0 = 0.75$
$c^- = 0.65$
$D_1 = 0.50$
$D_2 = 0.25$

The conditions for strong and weak outranking can then be outlined:

a) for Strong Outranking: aS^Fb

- $c(a,b) \geq 0.85$,
- $g_j(a) - g_j(b) \leq 0.50$, and
- $\dfrac{P^+(a,b)}{P^-(a,b)} \geq 1$

and / or

- $c(a,b) \geq 0.75$,
- $g_j(a) - g_j(b) \leq 0.25$, and
- $\dfrac{P^+(a,b)}{P^-(a,b)} \geq 1$

b) Weak Outranking: aS^fb

- $c(a,b) \geq 0.65$,
- $g_j(a) - g_j(b) \leq 0.50$, and
- $\dfrac{P^+(a,b)}{P^-(a,b)} \geq 1$

On the basis of these thresholds, the following outranking relationships were derived:

	MI	LA1	LA2	HI	SP
MI	-				
LA1	S^F	-	S^F	S^F	
LA2	S^F		-	S^F	
HI				-	
SP	S^F			S^F	-

Table 5-7 - Outranking Relationships for Weighting System 1

Direct Ranking Procedure
The results from the direct ranking procedure is summarised in the following table:

Step l	Y_l	D	U	B	A_l	$r_{1(l+1)}$
0	MI, LA1, LA2, HI, SP	LA1, SP	-	-	LA1, SP	1
1	MI, LA2, HI	LA2	-	-	LA2	2
2	MI, HI	MI, HI	-	-	MI, HI	3

Table 5-8 - Summary of Direct Ranking Procedure

The ranking from the direct procedure is as follows:

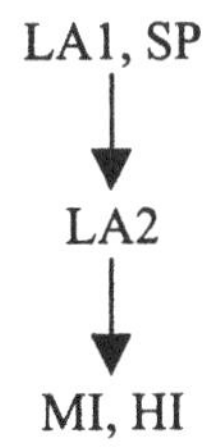

Inverse Ranking Procedure
Firstly, the outranking relationships given in Table 5-7 are inverted as follows:

	MI	LA1	LA2	HI	SP
MI	-	S^F	S^F		S^F
LA1		-			
LA2		S^F	-		
HI		S^F	S^F	-	S^F
SP					-

Table 5-9 - Summary of Inverted Outranking Relationships

These inverted relationships are then used as before to produce the ranking $r'_{2(l+1)}$ as follows:

Step l	Y_l	D	U	B	A_l	$r_{1(l+1)}$
0	MI, LA1, LA2, HI, SP	MI, HI	-	-	MI, HI	1
1	LA1, LA2, SP	LA2 SP	-	-	LA2, SP	2
2	LA1	LA1	-	-	LA1	3

Table 5-10 - Unadjusted Inverse Rankings

The adjusted inverse rankings $r_{2(l+1)}$ are estimated using the following equation:

$$r_{2(l+1)} = 1 + r'_{max} - r'_{2(l+1)}$$

Given that r'_{max}, the number of ranking positions in the inverse ranking procedure, equals 3:

$$r_{2(l+1)} = 4 - r'_{2(l+1)}$$

The adjusted inverse rankings then be computed as follows:

A_l	$r'_{2(l+1)}$	$r_{2(l+1)}$
MI, HI	1	3
LA2, SP	2	2
LA1	3	1

Table 5-11 - Adjusted Rankings

The ranking from the inverse procedure is:

Overall Ranking From Weighting System 1

The overall result from Weighting System 1 is therefore:

Sensitivity Analysis

The procedure was repeated for the following five different combinations of concordance and discordance thresholds:

	C^+	C^0	C^-	D_1	D_2
TEST 1	0.95	0.85	0.75	0.50	0.25
TEST 2	0.75	0.65	0.55	0.50	0.25
TEST 3	0.95	0.85	0.75	0.25	0.00
TEST 4	0.75	0.65	0.55	0.25	0.00
TEST 5	0.85	0.75	0.65	0.25	0.00

Table 5-12 - Threshold Variations

All of the five tests result in the same ranking as that shown above for the basic concordance and discordance thresholds.

Weighting System No. 2

Introduction

Using the lower quartile scores for Water quality and Marine Impacts, upper quartile scores for the other environmental effects, with cost remaining at its median score, the following normalised weighting system is derived:

Criterion	Vis	Amt	Bird	Wat	Mar	Con	Cst
System 2 Weighting	0.131	0.164	0.189	0.221	0.164	0.082	0.059

Table 5-13 - System No. 2 Weights

The resulting revised Concordance matrix is as follows:

	MI	**LA1**	**LA2**	**HI**	**SP**
MI	-	0.28	0.28	0.821	0.598
LA1	0.951	-	1	1	0.787
LA2	0.951	0.869	-	1	0.787
HI	0.951	0.221	0.221	-	0.598
SP	0.846	0.469	0.469	0.846	-

Table 5-14 - Concordance Matrix for Weighting System 2

The Discordance Matrix is identical to that shown in Table 5.6

The Concordance and Discordance thresholds determining the strong and weak outranking relationships are the same as for Weighting System 1, i.e.:

$c^+ = 0.85$
$c^0 = 0.75$

c^{-} = 0.65
D_1 = 0.50
D_2 = 0.25

On the basis of these thresholds, the following outranking relationships were derived:

	MI	**LA1**	**LA2**	**HI**	**SP**
MI	-				
LA1	S^F	-	S^F	S^F	S^f
LA2	S^F		-	S^F	
HI				-	
SP	S^F			S^F	-

Table 5-15 - Outranking Relationships for Weighting System 2

Direct Ranking Procedure
The results from the direct ranking procedure is summarised in the following table:

Step l	Y_l	D	U	B	A_l	$r_{l(l+1)}$
0	MI, LA1, LA2, HI, SP	LA1, SP	LA1, SP	LA1	LA1	1
1	SP, MI, LA2, HI	LA2, SP	-	-	LA2, SP	2
2	MI, HI	MI, HI	-	-	MI, HI	3

Table 5-16- Summary of Direct Ranking Procedure

The ranking from the direct procedure is as follows:

Inverse Ranking Procedure
Firstly, the outranking relationships given in Table 5-15 are inverted as follows:

	MI	**LA1**	**LA2**	**HI**	**SP**
MI	-	S^F	S^F		S^F
LA1		-			
LA2		S^F	-		
HI		S^F	S^F	-	S^F
SP		S^f			-

Table 5-17 - Summary of Inverted Outranking Relationships

These inverted relationships yield a ranking identical to that obtained by the direct ranking procedure immediately above.

The overall result from Weighting System 2 is therefore:

Sensitivity Analysis
Within Weighting System 2, the ELECTRE II Method was repeated for the same five combinations of concordance and discordance thresholds as illustrated in Table 5-12. For Tests 1 and 2, the ranking remains the same as for the base case. For Tests 3, 4 and 5, the ranking reverts to the base case for Weighting System 1 with SP moving above LA2 in the overall ranking.

Weighting System No. 3

Using the lower quartile scores for all environmental impacts, with the cost impact set at its upper quartile value to emphasise its importance, the following normalised weighting system is derived:

Criterion	Vis	Amt	Bird	Wat	Mar	Con	Cst
System 3 Weighting	0.071	0.04	0.081	0.273	0.2	0.040	0.293

Table 5-18 - System No. 3 Weights

The resulting revised Concordance matrix for Weighting System 3 is as follows:

	MI	**LA1**	**LA2**	**HI**	**SP**
MI	-	0.566	0.566	0.92	0.425
LA1	0.708	-	1	1	0.506
LA2	0.708	0.667	-	1	0.506
HI	0.708	0.273	0.273	-	0.425
SP	0.96	0.647	0.647	0.96	-

Table 5-19 - Concordance Matrix for Weighting System 3

The Discordance Matrix is identical to that shown in Table 5-6

The Concordance and Discordance thresholds determining the strong and weak outranking relationships are the same as for Weighting System 1, i.e.:

$c^+ = 0.85$
$c^0 = 0.75$
$c^- = 0.65$
$D_1 = 0.50$
$D_2 = 0.25$

On the basis of these thresholds, the following outranking relationships were derived:

	MI	**LA1**	**LA2**	**HI**	**SP**
MI	-			S^F	
LA1	S^f	-	S^F	S^F	
LA2	S^f		-	S^F	
HI				-	
SP	S^F			S^F	-

Table 5-20- Outranking Relationships for Weighting System 3

Direct Ranking Procedure
The results from the direct ranking procedure is summarised in the following table:

Step l	Y_l	D	U	B	A_l	$r_{l(l+1)}$
0	MI, LA1, LA2, HI, SP	LA1, SP	-		LA1, SP	1
1	MI, LA2, HI	LA2, MI	LA2, MI	LA2	LA2	2
2	MI, HI	MI	-	-	MI	3
4	HI	HI	-	-	HI	4

Table 5-21- Summary of Direct Ranking Procedure

The ranking from the direct procedure is as follows:

Inverse Ranking Procedure

Firstly, the outranking relationships given in Table 5-20 are inverted as follows:

	MI	LA1	LA2	HI	SP
MI	-	S^f	S^f		S^F
LA1		-			
LA2		S^F	-		
HI	S^F	S^F	S^F	-	S^F
SP					-

Table 5-22 - Summary of Inverted Outranking Relationships

These inverted relationships are then used as before to produce the ranking $r'_{2(l+1)}$ as follows:

Step l	Y_l	D	U	B	A_l	$r_{1(l+1)}$
0	MI, LA1, LA2, HI, SP	HI	-	-	HI	1
1	LA1, LA2, SP, MI	LA2, MI	LA2, MI	MI	MI	2
2	LA1, LA2, SP	LA2, SP	-	-	LA2, SP	3
4	LA1	LA1	-	-	LA1	4

Table 5-23 - Unadjusted Inverse Rankings

The adjusted inverse rankings $r_{2(l+1)}$ are estimated using the following equation:

$$r_{2(l+1)} = 1 + r'_{max} - r'_{2(l+1)}$$

Given that r'_{max}, the number of ranking positions in the inverse ranking procedure, equals 3:

$$r_{2(l+1)} = 5 - r'_{2(l+1)}$$

The adjusted inverse rankings then be computed as follows:

A_l	$r'_{2(l+1)}$	$r_{2(l+1)}$
HI	1	4
MI	2	3
LA2, SP	3	2
LA1	4	1

Table 5-24 - Adjusted Rankings

The ranking from the inverse procedure is:

The overall result from Weighting System 3 is therefore:

Sensitivity Analysis

Within Weighting System 3 the ELECTRE II Method was repeated for the same five combinations of concordance and discordance thresholds as illustrated in Table 5-12.

All tests emphasise HI as being the weakest option. For this set of weightings, LA1 is only marginally better than SP. Indeed, sensitivity tests 1 and 3, which raise the concordance thresholds over their base values, result in the two best options becoming indistinguishable.

For the other sensitivity tests, the ranking shows no significant variation from the base case rankings.

5.5 Overall Conclusion

The analysis shows the preferred options are Lough Atalia (Shallow Outfall) and South Park. For most weighting combinations, Lough Atalia is the better of the two, but they become indistinguishable as the importance of the cost option is emphasised.

A report commissioned by the European Commission subsequent to their consideration of the Environmental Impact Statement, and prior to a final funding decision being made, recommended Lough Atalia (shallow outfall) as their preferred option. This conflicted with the wishes of the local city council who finally chose a modified version of the Mutton Island proposal - an option ranked fourth out of five in the above analysis.

5.6 References

Canter, L. (1995) *Environmental Impact Assessment*. Mc Graw Hill, New York, "2nd Edition.

Casey, A.L. and Cronin, M.C. (1996) 'Environmental Impact Assessment and Decision Making in Civil Engineering'. Final Year Thesis, B.E. (Civil), University College Dublin (unpublished).

Department of Transport (1982) *Manual of Environmental Appraisal*. HMSO, London

Hokkanen, J. and Salminen, P (1997) Choosing a solid waste management system using multicriteria decision analysis'. *European Journal of Operational Research*, Vol. 98, p19-36, Elsevier B.V.

McCarthy, P.H. (1992) *Environmental Impact Statement, Galway Drainage Scheme*. Galway Corporation.

Mouchel, L.G. (1994) *Review of Alternative Sewerage and Wastewater Treatment Facilities for Galway City*. Report commissioned by the Save Galway Bay Group, January.

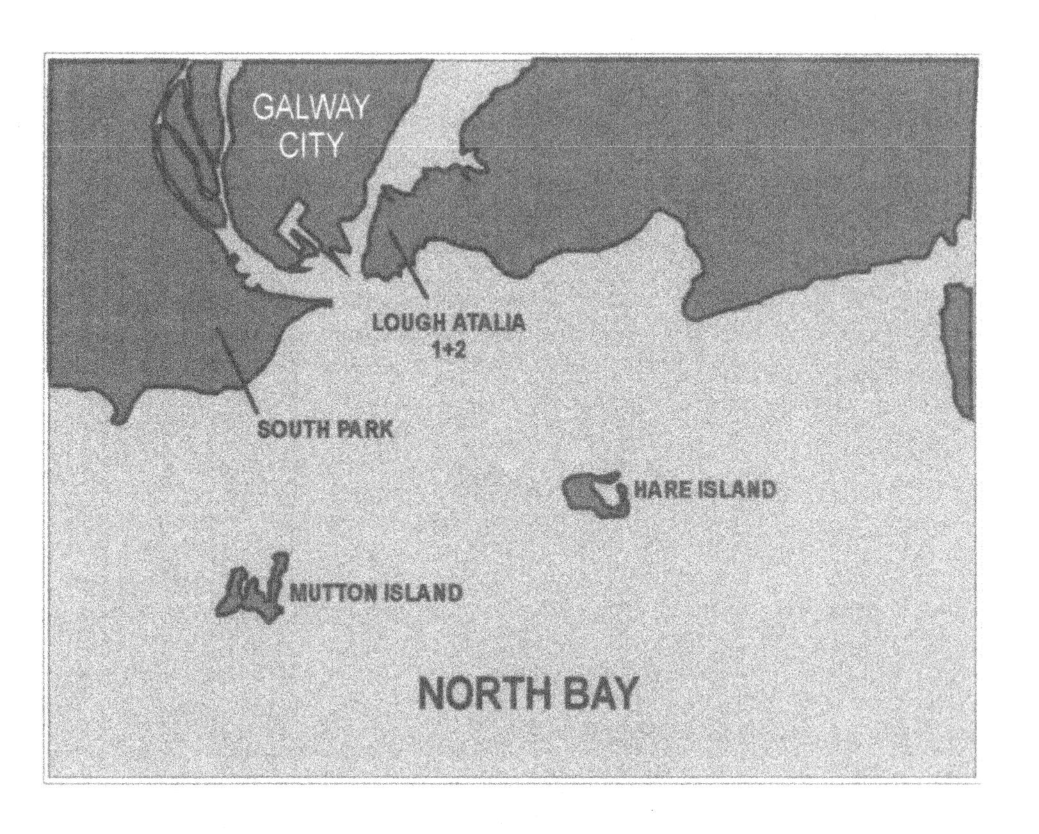
GALWAY CITY
LOUGH ATALIA 1+2
SOUTH PARK
HARE ISLAND
MUTTON ISLAND
NORTH BAY

Fig.5.1

6 CASE STUDY 2
Choosing the Best Waste Incineration Strategy for the Eastern Switzerland Region

6.1 Introduction

Switzerland is a confederation of 23 states, called cantons, three of which are sub-divided into half-cantons for administrative purposes. The powers of decision making are usually organised from the bottom upwards. In accordance with the 'subsidiarity principle', the Confederation only assumes responsibility for duties which the cantons are unable to carry out themselves. Thus the 26 cantons which make up the Federal State have a certain degree of sovereignty, and possess their own political institutions. Collegial government and therefore compromise are crucial factors in the political culture of the country. A number of years ago, cantons began, in co-operation with localities in other countries, to form *regions* which are primarity concerned with economic and infrastructural issues such as the siting of waste management plants. These regions as yet possess no political institutions of their own. One such region, Eastern Switzerland, comprising Schaffhausen, Uri, Zug, Zurich, Appenzell (divided in two), St Gallen, Thurgau, Glarus, Schwylz, Graubunden and Ticino, plus the Principality of Liechenstein, is the location for the case study outlined below.

Each of the 26 cantons has legislation enabling them to construct waste management facilities needed to meet environmental standards. Past reluctance of cantons to treat solid waste from neighbouring areas has led to a substantial overcapacity of waste incinerators in the Eastern Switzerland region. Given that, under Article 24 septies of the Federal Constitution, the Confederation is responsible, in legislative terms, for the protection of the 'human person and its natural environment against harmful acts or nuisances', the Federal Agency for the Environment invited a group of cantons from the Eastern Switzerland region to constitute a working group / expert group to decide on the **optimum waste strategy for the region** in the future. Federal law on the protection of the environment dictates that the cantons execute environmental measures such as the planning and construction of waste incineration or Waste-to-Energy (WTE) plants. Hence the need for compromise between federal and state interests.

For planning purposes, the region is divided into four zones plus the canton of Ticino as follows:

Zone 1 - The cantons of Shaffhausen, Uri, Zug and Zurich, containing 6 existing WTE Plants, with a total authorised capacity of 715 Ktons/Year.

Zone 2 - The two half cantons of Appenzell, part of St Gallen and Thurgau, containing 3 existing WTE Plants, with a total authorised capacity of 265 Ktons/Year.

Zone 3 - The cantons of Glarus, St Gallen (Part) and Schwartz, together with the Principality of Liechtenstein, containing one WTE Plant, with a total authorised capacity of 267 Ktons/Year

Zone 4 - The canton of Graubunden, containing 1 existing WTE Plant, with a total authorised capacity of 49 Ktons/Year

The canton of Ticino has no WTE Plant, but is heavily committed to the construction of a new installation in the near future.

6.2 The Strategic Options

The strategic options to be assessed were broken down into four broad categories:

Strategy 1 - Contruction of new WTE Plants (2 options developed, 1.1 and 1.2)

Strategy 2 - Maximise the transporting of waste between zones (4 options developed, 2.1, 2.2, 2.3 and 2.4)

Strategy 3 - Decentralising of waste facilities (2 options developed, 3.1 and 3.2)

Strategy 4 - Integrating / Merging of zones (3 options developed, 4.1, 4.2 and 4.3)

Each of the strategic options can be detailed as followed as follows:

Option S1.1

The construction of two new WTE Plants in Ticino and Graubunden, the maintenance of the WTE Plants in Zones 1, 2 and 3 at the maximum of their existing capacities, with no waste to be imported for treatment from other states

Option S1.2

The same basic strategy as S1.1, but with a provision for the treatment of waste imported from Austria and Germany

Option S2.1
The reduction of the capacites of WTE Plants in Zones 1 and 2, with no new plants to be constructed in Ticino and Graubunden, with waste from Ticino to be transported to Zones 1 and 2 for treatment and waste from Graubunden to be transported to Zone 3.

Option S2.2
The same basic strategy as S2.1, but with the provision that waste originating in Ticino should be transported for treatment to Zone 1 and waste from Graubunden to Zone 3.

Option S2.3
The same basic strategy as S2.2, but with the provision that waste from Germany be transported for treatment to Zones 1 and 2.

Option S2.4
The same basic strategy as S2.3, but with the provision that waste from Austria be transported to Zones 1 and 3.

Option S3.1
The capacities in Zones 1, 2 and 3 to be reduced, with two new plants to be constructed, one in Ticino, one in Graubunden, and 5 Ktons/year of waste to be transported for treatment from Zone 4 to Ticino.

Option S3.2
The capacities in Zone 1 to be reduced, with one new plant to be constructed in Ticino, one in Graubunden, and 5 Ktons/year of waste to be transported for treatment from Zone 4 to Ticino, plus the importation of waste from Austria and Germany to be disposed of in Zones 1 to 3.

Option S4.1
Zones 1 and 2 to remain in their existing form, with Zones 3 and 4 to be merged, and waste to be transported from Graubunden to Zone 3 for disposal. One new plant to be constructed in Ticino, with 5 Ktons/Year to be transported for disposal from Zone 4 to Ticino. Capacity of WTE Plants in Zones 1, 2 and 3 to be reduced.

Option S4.2
The same basic strategy as S4.1, but waste to be imported from Germany to Zones 1 and 2 for disposal.

Option S4.3
The same basic strategy as S4.2, plus waste to be imported from Austria to Zone 1 for disposal.

6.3 Decision Criteria

Each of the 11 strategic options were assessed on the following 11 environmental, economic, technical and political criteria:

Category	Criterion	Unit
Environmental	Waste Transportation Level (C1.1)	Ktons*km /Year
	Energy Levels Produced (C1.2)	Thermic GWh/Year
	Impact of Gas Emissions (C1.3)	Pop*Mtons Nox/Year
Economic	Avg. Treatment Costs within Region (C2.1)	SF/Ton
	Cost variability within Region (C2.2)	%
Technical	Adaptability to increases in waste production (C3.1)	Ktons/Year
	Adaptability to decreases in waste production (C3.2)	Ktons/Year
	Overcapacity in region by 2010 (C3.3)	%
Political	Opposition to new WTE Plants (C4.1)	score (0 to 1.5)
	Opposition to importing waste from other States (C4.2)	score (0 to 1.5)
	Reliability of supply of foreign waste for disposal (C4.3)	rank (1 to 11)

Table 6-1 - Evaluation Criteria

Let us look at the criteria in more detail

Environmental Criteria (C1)

Distance of Waste Transportation (C1.1)
This criterion reflects both the quantities of waste being transported, together with the distances being travelled. The unit of measurement is Kilotons of waste multiplied by distance travelled in Kilometres per year (Kt*km/Yr).

Energy Use (C1.2)
This criterion is an indicator of the energy actually used within the relevant regions, measured in Gigawatt hours per year (GWh/y).

Impact of Gas Emissions (C1.3)
The estimate for this criterion is obtained by multiplying the quantity of Nitrogen Oxides produced by an incinerator in Megatons per year by the population living within 2 Kilometres of the plant in question (Pop*MtNox/Yr).

Economic Criteria (C2)

Average Treatment Cost per Region (C2.1)
The treatment cost refers to the projected cost of incineration in the year 2010 in Swiss Francs per ton of waste incinerated. It includes all transportation, incineration

and energy production costs, minus the revenue from the sale of heat and electricity resulting from the process itself.

Uniformity of Treatment Costs (C2.2)
The cost of treatment for each of the zones in the study is estimated. For each of the options, this criterion gives the ratio between the most and least expensive zones relevant to that particular strategy. It reflects the unease that the public might feel if, within any of the strategic options outlined, large differences in cost should occur between the zones. It is expressed in terms of the ratio between the highest and lowest zonal cost for a given option.

Technical / Flexibility Criteria (C3)

Adaptability to Possible Increases in Waste Production (C3.1)
This criterion measures the flexibility or adaptability of a given option to possible increases in waste production over the next 15 years. The criterion details five alternative methods of dealing with this increase, each weighted in terms of their effectiveness in dealing with the increase:

Method for coping with increase	**Weighting factor**
Dealt with by means of:	
M1. Utilising any existing overcapacity	1.0
M2. Postponing the proposed closure of an incineration plant	0.8
M3. Ceasing the importation of waste	0.5
M4. Constructing a new incineration plant in Graubunden	0.2
M5. Constructing a new incineration plant in Ticino	0.1

Table 6-2 - Measures to deal with increases in waste production and their relative effectiveness

The criterion, measured in Kilotons per year, is estimated using the following index:

$$C3.1\ [Kt/y] = 1.0*M1[Kt/y] + 0.8*M2[Kt/y] + 0.5*M3[Kt/y] + 0.2*M4[Kt/y] + 0.1*M5[Kt/y]$$

Adaptability to Possible Decreases in Waste Production (C3.2)
In contrast to criterion C3.1, this criterion measures the flexibility or adaptability of a given option to possible decreases in waste production over the next 15 years. Here, the criterion details three alternative methods of dealing with any possible decrease, each weighted in terms of their effectiveness in dealing with the event:

Method for coping with decrease	**Weighting factor**
Dealt with by means of:	
M1. Increasing the importation of waste	1.0
M2. The closure of more treatment lines	0.8
M3. The closure of Incineration Plants	0.4

Table 6-3 - Measures to deal with decreases in waste production
and their relative effectiveness

The criterion, again measured in Kilotons per year, is estimated using the following index:

C3.2 [Kt/y] = 1.0*M1[Kt/y] + 0.8*M2[Kt/y] + 0.4*M3[Kt/y]

In both cases, the higher the adaptability score, the better the option

Overcapacity (C3.3)
The expert group considered an overcapacity within any option of 12% to be sufficient to deal with any possible seasonal fluctuations up to the year 2010. Any production over this limit is therefore deemed excess to requirement and is estimated as follows:

C3.3 [%] = (100 * total capacity[Kt/y] / total waste treated [Kt/y]) - 112

The higher the level of net overcapacity, the less attractive the option in question.

Political Criteria (C4)

Opposition of Local Pressure Groups (C4.1)
The working group evaluated each option on a three point scale, 0, 1.0 and 1.5 based on the level of opposition, if any, local groups would have to the new waste management strategy. The working group judged that opposition would be low if the option did not entail the construction of a new incineration plant, that opposition would be substantial if such construction was envisaged, and that oposition would be greater again if both the construction of a new plant and the upgrading of existing facilities was planned. This scale can be detailed as follows:

Level of Opposition	**Opposition Score**
M1. If **no** new incineration plant is planned	0.0
M2. If a new plant **is** planned	1.0
M3. If a new incineration plant **plus** the enlargment of existing facilities is Planned	1.5

Table 6-4 - Opposition of Local Action Groups

The lower the score, the less the perceived local opposition to the option in question, and the greater its attraction.

Opposition to the Importation of Foreign Waste (C4.2)
This criterion was scored by the working group on an 8-point scale, with the lowest score being the most preferable. The score reflects the level of political opposition to different levels of importation, with the most preferable politically being no importation, and the least preferable being the importation of waste into East Switzerland from all potential Cantons / Countries - Graubunden, Ticino, Germany and Austria. The scale is detailed as follows:

IMPORT STRATEGY	**SCORE**
No importing of waste	1
Importing of waste from Graubunden only	2
Importing of waste from Graubunden and Ticino	3
Importing of waste from Graubunden and Germany	4
Importing of waste from Graubunden, Ticino and Germany	5
Importing of waste from Austria and Germany	6
Importing of waste from Graubunden, Germany and Austria	7
Importing of waste from Graubunden, Ticino, Germany and Austria	8

Table 6-5 - Preference for Different Importation Strategies

Dependancy on the Supply of Imported waste (C4.3)
The working group scored this criterion on an 7-point scale, with the lowest score being the most preferable. The score reflects the level of risk attached to depending on a steady supply of imported waste. The criterion deems no dependancy on imported waste to be the most preferable state. Next preferable is the case where only Graubunden is required by its Administration to export its waste to East Switzerland. The least preferable state is deemed to exist where all neighbouring Cantons and States - Graubunden, Ticino, Germany and Austria - are required by their respective Administrations to export their waste to East Switzerland. The scale is based on the assumption that, because this supply depends on the policy decisions of administrations outside the control of Eastern Switzerland, there is therefore an element of risk and uncertainty attached to it. The scale is detailed as follows:

DEPENDANCY ON IMPORTED WASTE	SCORE
No importing of waste	1
Graubunden is required to export its waste to East Switzerland	2
Ticino is required to export its waste to East Switzeralnd	3
Germany is required to export its waste	4
Ticino and Germany are required to export their waste to ES	5
Austria and Germany are required to export their waste to ES	6
Ticino, Germany and Austria are required to export their waste	7

Table 6-6 - Dependence on Imported Waste

6.4 The Performance matrix

A summary of the performance of each option on all 11 criteria is as follows:

Options	Criteria										
	C1.1	C1.2	C1.3	C2.1	C2.2	C3.1	C3.2	C3.3	C4.1	C4.2	C4.3
	Kton. km	Gwt/year	Gt/y* cap	SFr/ Ton	-	Kton/ Year	Kton/ Year	%	Score	Rank	Rank
Dir. Of Pref.	↓	↑	↓	↓	↓	↑	↑	↓	↓	↓	↓
1.1	*125*	*866*	*9.81*	*218*	*1.41*	*542*	*483*	*23*	*1.5*	*1*	*1*
1.2	*11980*	*900*	*11.45*	*189*	*1.45*	*452*	*303*	*12*	*1.5*	*6*	*6*
2.1	*31054*	*883*	*9.86*	*172*	*1.82*	*341*	*311*	*0*	*0*	*3*	*3*
2.2	*28219*	*840*	*10.38*	*171*	*1.95*	*339*	*318*	*0*	*0*	*3*	*3*
2.3	*31579*	*903*	*10.74*	*165*	*1.7*	*312*	*281*	*0*	*0*	*5*	*5*
2.4	*39364*	*922*	*13.87*	*167*	*1.65*	*287*	*269*	*0*	*0*	*8*	*7*
3.1	*125*	*769*	*9.33*	*182*	*1.64*	*458*	*180*	*0*	*1.5*	*1*	*1*
3.2	*8075*	*896*	*9.82*	*172*	*1.7*	*408*	*121*	*0*	*1.5*	*6*	*6*
4.1	*3089*	*770*	*9.39*	*177*	*1.9*	*430*	*228*	*0*	*1*	*2*	*2*
4.2	*6449*	*766*	*7.22*	*172*	*1.65*	*401*	*157*	*0*	*1*	*4*	*4*
4.3	*12074*	*897*	*10.61*	*169*	*1.65*	*378*	*162*	*0*	*1*	*7*	*6*

Table 6-7 - Performance Matrix for the 11 Strategic Options

Data for the eight quantitative criteria was formulated by independent consultants employed by the group. The data for the three qualitative criteria were established by the negotiating group itself.

6.5 The Working Group

In Switzerland, the commune is the basic unit of local government. There are 3018 in all. Like the Federal Government and the cantons, each has its own government and parliament. Each of the incineration plants is operated by intercommunal organisations, which are independent of both the Federal Office for the Environment and the cantons. Given the necessity to balance all three levels of administration, the endeavour to arrive at compromise in issues such as this is at the heart of Swiss politics. Thus, while the Confederation is required to co-ordinate the environmental issues central to the planing and construction of waste disposal installations such as WTE Plants, the cantons are responsible for the execution of these federal prescriptions, while the intercommunal associations actually operate these plants. To achieve the necessary compromise in this instance, the FAE invited the cantons and intercommunal associations from Eastern Switzerland to constitute a working group that would agree on an optimum strategy for waste incineration in the region. For the purpose of decision making, it was decided to designate the following voting system. Each voter in the group comprised one canton representative plus one representative from the relevant intercommunal associations (The Canton of Ticino did not want to participate for political reasons, and was therefor requested to send a non-voting observer).

The following was the composition and voting rights of each constituent group:

Zone	Voting Power	Agency Representatives	Representatives of Intercommunal Associations
1	1	1 from Canton Zurich	1 from 6 associations
2	2	2 from canton St Gallen	1 from 2 assocations
		1 from Canton Thurgau	1 from 1 association
3	1	1 from canton Glarus	1 from 1 association
4	1	1 from canton Graubunden	1 from 1 association
-	1	2 from FAE	-

Table 6-8- Actors in the Negotiating Group

6.6 Criterion Weightings

The task of fixing weights is a crucial step in the operation of the ELECTRE III model. It was essential that all members of the negotiating group had the opportunity to interpret weightings of their own, which could then be input to the model. It was decided that each of the above 6 groupings, from the three different levels of administration in Switzerland, would have an equal voice in the decision. To achieve this, the group decided that each voting unit referred to in Table 6.8 would be entitled to input its own set of importance weightings into the model. The two main weighting techniques detailed in chapter 4 – The Revised Simos Technique (Roy and Figueira, 1998) and The Resistence-to-Change Grid (Rogers and Bruen, 1998) – were used by all voting units to derive their individual set of weights. In all cases, the differences in the weights resulting from the two procedures were minor. All scores were normalised. The resulting importance weights derived are shown in Table 6.9 below.

Criterion	Zurich	Glarus	StGallen	Graubund	Thurgau	FAE	Average Weights
C.1.1	0.047	0.128	0.121	**0.173**	0.107	0.16	0.13
C.1.2	**0.122**	0.035	0.05	0.079	0.06	0.033	0.06
C.1.3	0.2	**0.106**	0.045	0.071	0.019	0.033	0.05
C.2.1	0.116	0.177	0.138	**0.193**	**0.193**	0.097	0.15
C.2.2	0.087	0.09	0.065	0.12	**0.124**	0.097	0.10
C.3.1	0.13	0.012	0.098	0.053	**0.162**	0.16	0.10
C.3.2	0.144	0.029	0.112	0.053	**0.162**	0.097	0.10
C.3.3	0.068	0.102	0.093	0.12	0.068	**0.16**	0.10
C.4.1	0.102	**0.151**	0.117	0.013	0.067	0.033	0.08
C.4.2	0.082	**0.093**	0.084	0.013	0.019	0.033	0.05
C.4.3	0.082	0.077	0.079	**0.112**	0.019	0.097	0.08
Total	1	1	1	1	1	1	1

Table 6-9 - Importance Weightings for Different Actors and Overall Averages

The highest weighting for each criterion is indicated in bold within Table 6.9. Except for C2.1 and C2.2, strong divergences exist in the perceived importances of the criteria. In the case of C2.1 and C2.2, the ratio of maximum to minimum weighting is approximately two. In the case of others such as C3.1 and C4.1, this ratio is in excess of 10.

6.7 Criterion Thresholds

The pseudo-criterion used in ELECTRE III requires specified indifference, preference and veto thresholds. Fixing the thresholds involves not only the estimation of error in a physical sense, but also a significant subjective input by the decision-makers themselves (Rogers and Bruen 1998). Maystre et al. (1994) interpreted the indifference threshold q as the minimum margin of imprecision associated with a given criterion and the preference threshold p as the maximum margin of error associated with the criterion in question. In order to estimate values of p and q for the purposes of the case study, independent consultants provided the basic information regarding the level of imprecision associated with each of the criterion valuations. This enabled the negotiating group to select indifference thresholds, directly related in each case to the minimum level of imprecision that could be attributed to each set of criterion valuations. The negotiating group then set the preference threshold in most cases as twice the value of the preference threshold.

The veto threshold v characterises the situation where a discordant criterion can, on its own, exert a veto on an entire outranking relationship. It is must at least be set equal to the preference threshold p, and is usually set at three times this value. The ELECTRE III model can work without the assignment of veto thresholds to the criteria. In this case, the concordance matrix is set equal to the credibility matrix. The negotiating group decided that a veto threshold should be set for only three of the criteria – Transportation (C1.1), Average Treatment Cost (C2.1) and Overcapacity (C3.3). In these cases, the veto thresholds chosen by the group varied from just under four to exactly ten times the value of the preference threshold for the criterion in question.

The indifference, preference and, where appropriate, veto thresholds for each of criteria are given below in Table 6.10:

Criterion	Unit	Amplitude	Thresholds		
			Indiff.	Pref.	Veto
Transportation (C1.1)	Kton*km/Yr	125-39400	±1000	± 2000	± 20000
Energy Use (C1.2)	Gwt/year	766-922	10%	20%	-
Gas Emissions (C1.3)	Pop*Mt/year	7.22-13.87	10%	20%	-
Treatment Cost (C2.1)	SFr/Ton	165- 218	± 5	± 10	± 50
Uniformity (C2.2)	-	1.41-1.95	10%	20%	-
Adaptability to increases (C3.1)	Kton/Year	287-542	10%	20%	-
Adaptability to decreases (C3.2)	Kton/Year	121-483	10%	20%	-
Overcapacity (C3.3)	%	0-23	± 2	± 4	± 15
Local Opposition (C4.1)	Score	0.0-1.5	0.2	0.4	-
Opposition to foreign waste (C4.2)	Rank	1-8	± 0	± 1	-
Dependency on foreign waste (C4.3)	Rank	1-7	± 0	± 1	-

Table 6-10 – Criterion Threshold Values

It was decided by the negotiating group not to use the veto thresholds within the baseline run of the ELECTRE III model.

6.8 The Outranking Matrix

Information from Tables 6.7, 6.9 and 6.10 were combined within the ELECTRE III model to derive a Concordance Matrix for each of the six actors providing a set of weights. Since, in the baseline run of the model, veto thresholds were ignored, the matrix is identical to the Degree of Credibility Matrix. For demonstration purposes, we will concentrate on an analysis of the ranking which results from the weightings from one of the groups –the Federal Agency for the Environment (FAE). The Degree of Credibility of Outranking Matrix derived for this actor is as follows:

	S1.1	S1.2	S2.1	S2.2	S2.3	S2.4	S3.1	S3.2	S4.1	S4.2	S4.3
S1.1	1	0.74	0.71	0.71	0.71	0.71	0.74	0.74	0.71	0.68	0.71
S1.2	0.44	1	0.57	0.58	0.58	0.71	0.48	0.57	0.39	0.39	0.71
S2.1	0.36	0.58	1	0.84	0.96	1	0.55	0.69	0.55	0.69	0.83
S2.2	0.36	0.58	1	1	0.95	0.95	0.49	0.65	0.55	0.62	0.76
S2.3	0.38	0.63	0.86	0.68	1	1	0.54	0.68	0.54	0.52	0.68
S2.4	0.38	0.48	0.48	0.47	0.68	1	0.52	0.52	0.52	0.52	0.52
S3.1	0.72	0.86	0.76	0.77	0.75	0.74	1	0.88	0.87	0.84	0.848
S3.2	0.38	0.84	0.74	0.74	0.7	0.87	0.58	1	0.58	0.61	0.87
S4.1	0.35	0.78	0 .85	0.85	0.74	0.73	0.67	0.97	1	0.94	0.89
S4.2	0.4	0.81	0.7 2	0.74	0.81	0.84	0.6	0.98	0.61	1	0.98
S4.3	0.41	0.7	0.74	0.74	0.74	0.87	0.53	0.81	0.55	0.68	1

Table 6-11 – Degree of Credibility Matrix for FAE

Sample Outranking Calculations

Taking a pair of options S1.1 and S2.2:

	C1.1	C1.2	C1.3	C2.1	C2.2	C3.1	C3.2	C3.3	C4.1	C4.2	C4.3
S1.1	125	866	9.81	218	1.41	542	483	23	1.5	1	1
S2.2	28219	840	10.38	171	1.95	339	318	0	0	3	3

Let us examine the degree of credibility we can attach to their outranking of S1.1 over S2.2:

C(S1.1,S2.2)
Put a = S1.1, b = S2.2
(i) criterion C1.1 (preference in favour of smaller scores)
$g_{C1.1}(a) = 125$
$g_{C1.1}(b) = 28219$
since $g_{C1.1}(a) \leq g_{C1.1}(b)$, $c_{C1.1}(a,b) = 1$

(ii) criterion C1.2 (preference in favour of larger scores)
$g_{C1.2}(a) = 866$
$g_{C1.2}(b) = 840$
since $g_{C1.2}(a) \geq g_{C1.2}(b)$, $c_{C1.2}(a,b) = 1$

(iii) criterion C1.3 (preference in favour of smaller scores)
$g_{C1.3}(a) = 9.81$
$g_{C1.3}(b) = 10.38$
since $g_{C1.3}(a) \leq g_{C1.3}(b)$, $c_{C1.3}(a,b) = 1$

(iv) criterion C2.1 (preference in favour of smaller scores)
$g_{C2.1}(a) = 218$
$g_{C2.1}(b) = 171$
strict preference in favour of b requires that $g_{C2.1}(a) - g_{C2.1}(b) > p_{C2.1}[g_{C2.1}(a)]$
since $218 - 171 > 0.1*218$, $c_{C2.1}(a,b) = 0$

(v) criterion C2.2 (preference in favour of smaller scores)
$g_{C2.2}(a) = 1.41$
$g_{C2.2}(b) = 1.95$
since since $g_{C2.2}(a) \leq g_{C2.2}(b)$, $c_{C2.2}(a,b) = 1$

(vi) criterion C3.1 (preference in favour of larger scores)
$g_{C3.1}(a) = 542$
$g_{C3.1}(b) = 339$
since $g_{C3.1}(a) \geq g_{3.1}(b)$, $c_{C3.1}(a,b) = 1$

(vii) criterion C3.2 (preference in favour of larger scores)
$g_{C3.2}(a) = 483$
$g_{C3.2}(b) = 318$
since $g_{C3.2}(a) \geq g_{3.2}(b)$, $c_{C3.2}(a,b) = 1$

(viii) criterion C3.3 (preference in favour of smaller scores)
$g_{C3.3}(a) = 23$
$g_{C3.3}(b) = 0$
strict preference in favour of b requires that $g_{C3.3}(a) - g_{C3.3}(b) > p_{C3.3}[g_{C3.3}(a)]$
since $23 - 0 > 4$, $c_{C3.3}(a,b) = 0$
(ix) criterion C4.1 (preference in favour of smaller scores)
$g_{C4.1}(a) = 1.5$
$g_{C4.1}(b) = 0$
strict preference in favour of b requires that $g_{C4.1}(a) - g_{C4.1}(b) > p_{C4.1}[g_{C4.1}(a)]$
since $1.5 - 0 > 1$, $c_{C4.1}(a,b) = 0$

(x) criteria C4.2 (preference in favour of smaller scores)
$g_{C4.2}(a) = 1$
$g_{C4.2}(b) = 3$
since $g_{C4.2}(a) \leq g_{C4.2}(b)$, $c_{C4.2}(a,b) = 1$

(x) criteria C4.3 (preference in favour of smaller scores)
$g_{C4.3}(a) = 1$
$g_{C4.3}(b) = 3$
since $g_{C4.3}(a) \leq g_{C4.3}(b)$, $c_{C4.3}(a,b) = 1$

$$C(a,b) = \frac{1}{W} \sum_{j=1}^{n} w_j c_j(a,b)$$

where

$$W = \sum_{j=1}^{n} w_j$$

Since the weights given in Table 6-9 are normalised, *W* equals 1, thus C(a,b) can be denoted as follows:

$$C(a,b) = \sum_{j=1}^{n} w_j c_j(a,b)$$

Thus, by multiplying the concordance index calculated for each criterion by its normalised weighting score indicated in Table 6.9 as follows:

$$C(a,b) = 0.16*1 + 0.033*1 + 0.033*1 + 0.097*0 + 0.097*1 + 0.16*1 + 0.097*1 + 0.16*0 + 0.033*0 + 0.033*1 + 0.097*1 = 0.71$$

Therefore, since the veto threshold is muted, this value becomes the degree of credibility of outranking score for this pair of options, thus:
S(S1.1, S2.2) = 0.71 as indicated in Table 6.11

6.9 Exploiting the Outranking Relationships

Introduction

In order to devise a ranking of all options, two pre-orders must be constructed, descending and ascending.

Their construction requires a qualification score for each option to be estimated.

Firstly, let λ_0 equal to the the maximum value of S(a,b) in Table 6-11.
Therefore,

$$\lambda_0 = 1$$

A cut-off level of outranking λ_1 is defined as the largest outranking score which is just less than the maximum outranking score lessthe discrimination threshold, i.e.:

$$\lambda_1 = \max_{\{S(a,b) < \lambda_0 - s(\lambda_0)\}} S(a,b)$$

where $s(\lambda_0)$ is the discrimination threshold at the maximum level of outranking λ_0.

At this initial cut-off level , *a* will outrank *b* if S(a,b) is greater than the cut-off level, and S(a,b) exceeds S(b,a) by more than the discrimination threshold, i.e.:

aSb iff $S(a,b) > \lambda_1$ and $S(a,b) - S(b,a) > s(S(a,b))$

Within ELECTRE III, the discrimination threshold is usually set at the following:

$s(\lambda) = 0.3 - 0.15\lambda$

Downward Distillation

$A_0 = A$, $s(\lambda) = 0.3 - 0.15\lambda$

- 1st Distillation

Step 1

$l = 0$, $D_0 = A_0$; $\lambda_0 = 1$; $\lambda_0 - s(\lambda_0) = 0.85$; $\lambda_1 = 0.848$.

	S1.1	**S1.2**	**S2.1**	**S2.2**	**S2.3**	**S2.4**	**S3.1**	**S3.2**	**S4.1**	**S4.2**	**S4.3**
S	-	-	2.4	S2.1 S2.3 S2.4	S2.4	-	S1.2 S3.2 S4.1	S2.4	S2.1 S2.2 S3.2 S4.2 S4.3	S3.2 S4.3	S2.4
Strength	0	0	1	3	1	0	3	1	5	2	1
Weakn's	0	1	2	1	1	5	0	3	1	1	2
Qualif'n	0	-1	-1	2	0	-5	3	-2	4	1	-1

$C_1 = D_1 = \{S4.1\}$

$A_1 = A_0 \setminus C_1 = \{S1.1, S1.2, S2.1, S2.2, S2.3, S2.4, S3.1, S3.2, S4.2, S4.3\}$

- 2nd Distillation

Step 1

$l = 0$, $D_0 = A_1$; $\lambda_0 = 1$; $\lambda_0 - s(\lambda_0) = 0.85$; $\lambda_1 = 0.848$.

	S1.1	**S1.2**	**S2.1**	**S2.2**	**S2.3**	**S2.4**	**S3.1**	**S3.2**	**S4.2**	**S4.3**
S	-	-	2.4	S2.1 S2.3 S2.4	S2.4	-	S1.2 S3.2	S2.4	S3.2 S4.3	S2.4
Strength	0	0	1	3	1	0	2	1	2	1
Weakn's	0	1	1	0	1	5	0	2	0	1
Qualif'n	0	-1	0	3	0	-5	2	-1	2	0

$C_2 = D_1 = \{S2.2\}$

$A_2 = A_1 \setminus C_2 = \{S1.1, S1.2, S2.1, S2.3, S2.4, S3.1, S3.2, S4.2\ S4.3\}$

- 3rd Distillation

Step 1

$l = 0$, $D_0 = A_2$; $\lambda_0 = 1$; $\lambda_0 - s(\lambda_0) = 0.85$; $\lambda_1 = 0.848$.

	S1.1	S1.2	S2.1	S2.3	S2.4	S3.1	S3.2	S4.2	S4.3
S	-	-	2.4	S2.4	-	S1.2 S3.2	S2.4	S3.2 S4.3	2.4
Strength	0	0	1	1	0	2	1	2	1
Weakn's	0	1	0	0	4	0	2	0	1
Qualif'n	0	-1	1	1	-4	2	-1	2	0

$D_1 = \{S3.1, S4.2\}$

Step 2

$l = 1$, $D_1 = \{S3.1, S4.2\}$; $\lambda_1 = 0.848$; $\lambda_1 - s(\lambda_1) = 0.675$; $\lambda_2 = 0.53$.

	S3.1	S4.3
S	S4.3	-
Strength	1	0
Weakn's	0	1
Qualif'n	1	-1

S(S3.1, S4.3) = 0.848
S(S4.3, S3.1) = 0.53
s (**S**(S3.1, S4.3)) = (0.3 – (0.15*0.848)) = 0.173
Since **S**(S3.1, S4.3) > λ_2, **S**(S3.1, S4.3) - **S**(S4.3, S3.1) > s (**S**(S3.1, S4.3)),
S3.1 outranks S4.3

Therefore,
$D_2 = C_3 = \{S3.1\}$
$A_3 = A_2 \setminus C_3 = \{S1.1, S1.2, S2.1, S2.3, S2.4, S3.2, S4.2\ S4.3\}$

- 4th Distillation

Step 1

$l = 0$, $D_0 = A_3$; $\lambda_0 = 1$; $\lambda_0 - s(\lambda_0) = 0.85$; $\lambda_1 = 0.84$

	S1.1	S1.2	S2.1	S2.3	S2.4	S3.2	S4.2	S4.3
S	-	-	2.4	S2.4	-	S2.4	S3.2 S4.3	S2.4
Strength	0	0	1	1	0	1	2	1
Weakn's	0	0	0	0	4	1	0	1
Qualif'n	0	0	1	1	-4	0	2	0

$D_1 = C_4 = \{S4.2\}$
$A_4 = A_3 \setminus C_4 = \{S1.1, S1.2, S2.1, S2.3, S2.4, S3.2, S4.3\}$

- 5th Distillation

Step 1
$l = 0$, $D_0 = A_4$; $\lambda_0 = 1$; $\lambda_0 - s(\lambda_0) = 0.85$; $\lambda_1 = 0.84$

	S1.1	S1.2	S2.1	S2.3	S2.4	S3.2	S4.3
S	-	-	2.4	S2.4	-	S2.4	S2.4
Strength	0	0	1	1	0	1	1
Weakn's	0	0	0	0	4	0	0
Qualif'n	0	0	1	1	-4	1	1

$D_1 = \{S2.1, S2.3, S3.2, S4.3\}$

Step 2
$l = 1$, $D_1 = \{S2.1, S2.3, S3.2, S4.3\}$
since $S(a,b) - S(b,a) < 0.15 = \text{Min } s(\lambda)$, $\forall$ $(a,b), (b,a) \in D_1$

	S2.1	S2.3	S3.2	S4.3
S	-	-	-	-
Strength	0	0	0	0
Weaknes s	0	0	0	0
Qualif'n	0	0	0	0

$D_2 = C_5 = \{S2.1, S2.3, S3.2, S4.3\}$
$A_5 = A_4 \setminus C_5 = \{S1.1, S1.2, S2.4\}$

- 6th Distillation

Step 1
$l = 0$, $D_0 = A_5$; $\lambda_0 = 0.74$; $\lambda_0 - s(\lambda_0) = 0.55$; $\lambda_1 = 0.48$

	S1.1	**S1.2**	**S2.4**
S	S1.2	S2.4	-
	S2.4		
Strength	2	1	0
Weakn's	0	1	2
Qualif'n	2	0	-2

$D_1 = C_6 = \{S1.1\}$
$A_6 = A_5 \setminus C_6 = \{S1.2, S2.4\}$

- 7th Distillation

Step 1
$l = 0$, $D_0 = A_6$; $\lambda_0 = 0.71$; $\lambda_0 - s(\lambda_0) = 0.517$; $\lambda_1 = 0.48$

	S1.2	**S2.4**
S	S2.4	-
Strength	1	0
Weakn's		1
Qualif'n	1	-1

$D_1 = C_7 = \{S1.2\}$
$A_7 = A_6 \setminus C_7 = \{S2.4\}$
Since A_7 is a singleton,
$C_8 = \{S2.4\}$

The resultant ranking obtained from the downward distillation is as follows:

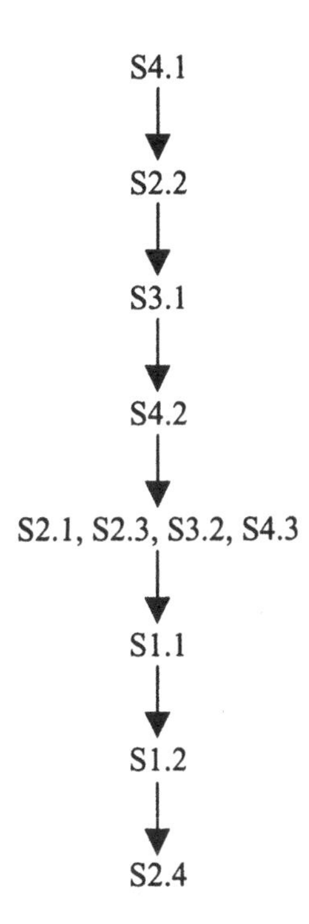

Upward Distillation

Using the same table as for the first downward distillation, but with qualification in this instance based on the least qualification score:

$A_0 = A$, $s(\lambda) = 0.3 - 0.15\lambda$

- 1st Distillation

Step 1
$l = 0$, $D_0 = A_0$; $\lambda_0 = 1$; $\lambda_0 - s(\lambda_0) = 0.85$; $\lambda_1 = 0.848$.
$C_1 = D_1 = \{S2.4\}$
$A_1 = A_0 \setminus C_1 = \{S1.1, S1.2, S2.1, S2.2, S2.3, S3.1, S3.2, S4.1, S4.2, S4.3\}$

- 2nd Distillation

Step 1
$l = 0$, $D_0 = A_1$; $\lambda_0 = 1$; $\lambda_0 - s(\lambda_0) = 0.85$; $\lambda_1 = 0.848$.

	S1.1	S1.2	S2.1	S2.2	S2.3	S3.1	S3.2	S4.1	S4.2	S4.3
S	-	-	-	S2.1 , S2.3	-	S1.2 S3.2 S4.1	-	S2.1 S2.2 S3.2 S4.2 S4.3	S3.2 S4.3	-
Strength	0	0	0	2	0	3	0	5	2	0
Weakn's	0	1	2	1	1	0	3	1	1	2
Qualif'n	0	-1	-2	1	-1	3	-3	4	1	-2

$C_2 = D_1 = \{S3.2\}$
$A_2 = A_1 \setminus C_2 = \{S1.1, S1.2, S2.1, S2.2, S2.3, S3.1, S4.1, S4.2, S4.3\}$

- 3rd Distillation

Step 1
$l = 0$, $D_0 = A_2$; $\lambda_0 = 1$; $\lambda_0 - s(\lambda_0) = 0.85$; $\lambda_1 = 0.848$.

	S1.1	S1.2	S2.1	S2.2	S2.3	S3.1	S4.1	S4.2	S4.3
S	-	-	-	S2.1 S2.3	-	S1.2 S4.1	S2.1 S2.2 S4.2 S4.3	S4.3	-
Strength	0	0	0	2	0	2	4	1	0
Weakn's	0	1	2	1	1	0	1	1	2
Qualif'n	0	-1	-2	1	-1	2	3	0	-2

$D_1 = \{S2.1, S4.3\}$

Step 2
$l = 1$, $D_1 = \{S2.1, S4.3\}$
since $\mathbf{S}(a,b) - \mathbf{S}(b,a) = 0.09 < 0.15 = \text{Min } s(\lambda)$
$C_3 = D_2 = \{S2.1, S4.3\}$
$A_3 = A_2 \setminus C_3 = \{S1.1, S1.2, S2.2, S2.3, S3.1, S4.1, S4.2\}$

- 4th Distillation

Step 1
$l = 0$, $D_0 = A_3$; $\lambda_0 = 0.95$; $\lambda_0 - s(\lambda_0) = 0.79$; $\lambda_1 = 0.78$.

	S1.1	S1.2	S2.2	S2.3	S3.1	S4.1	S4.2
S	-	-	S2.3	-	S1.2 S4.1 S4.2	S2.2 S4.2	S1.2 S2.3
Strength	0	0	1	0	3	2	2
Weakn's	0	2	1	2	0	1	2
Qualif'n	0	-2	0	-2	3	1	0

$D_1 = \{S1.2, S2.3\}$

Step 2
$l = 1$, $D_1 = \{S1.2, S2.3\}$
since $S(a,b) - S(b,a) = 0.05 < 0.15 = \text{Min } s(\lambda)$
$C_4 = D_2 = \{S1.2, S2.3\}$
$A_4 = A_3 \setminus C_4 = \{S1.1, S2.2, S3.1, S4.1, S4.2\}$

- 5th Distillation

Step 1
$l = 0$, $D_0 = A_4$; $\lambda_0 = 0.94$; $\lambda_0 - s(\lambda_0) = 0.78$; $\lambda_1 = 0.77$.

	S1.1	S2.2	S3.1	S4.1	S4.2
S	-	-	S4.1 S4.2	S2.2 S4.2	-
Strength	0	0	2	2	0
Weakn's	0	1	0	1	2
Qualif'n	0	-1	2	1	-2

$C_5 = D_1 = \{S4.2\}$
$A_5 = A_4 \setminus C_5 = \{S1.1, S2.2, S3.1, S4.1\}$

- 6th Distillation

Step 1
$l = 0$, $D_0 = A_5$; $\lambda_0 = 0.87$; $\lambda_0 - s(\lambda_0) = 0.70$; $\lambda_1 = 0.67$.

	S1.1	S2.2	S3.1	S4.1
S	S2.2 S4.1	-	S4.1 S2.2	S2.2
Strength	2	0	2	1
Weakn's	0	3	0	1
Qualif'n	2	-3	2	1

$C_6 = D_1 = \{S2.2\}$
$A_6 = A_5 \setminus C_6 = \{S1.1, S3.1, S4.1\}$

- 7th Distillation

Step 1

$l = 0$, $D_0 = A_6$; $\lambda_0 = 0.87$; $\lambda_0 - s(\lambda_0) = 0.70$; $\lambda_1 = 0.67$.

	S1.1	S3.1	S4.1
S	S4.1	S4.1	-
Strength	1	1	0
Weakn's	0	0	2
Qualif'n	1	1	-2

$C_7 = D_1 = \{S4.1\}$

$A_7 = A_6 \setminus C_7 = \{S1.1, S3.1\}$

- 8th Distillation

$l = 0$, $D_0 = \{S1.1, S3.1\}$

since $S(a,b) - S(b,a) = 0.02 < 0.15 = \text{Min } s(\lambda)$

the two options are inseparable

therefore

$C_8 = D_1 = \{S1.1. S3.1\}$

The resultant ranking obtained from the upward distillation is as follows:

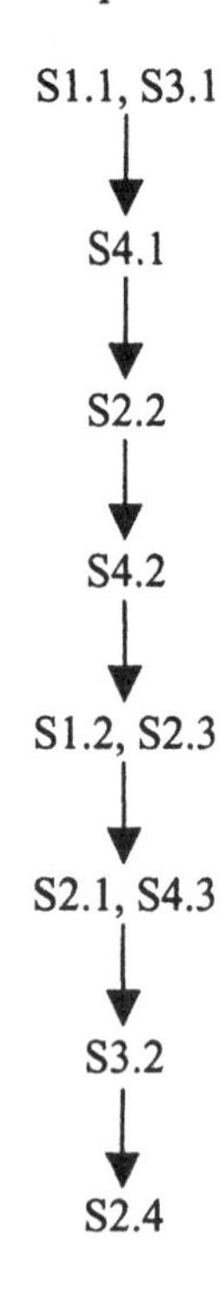

Final Ranking

The downward and upward distillations are combined to give the overall ranking for the base case situation as follows:

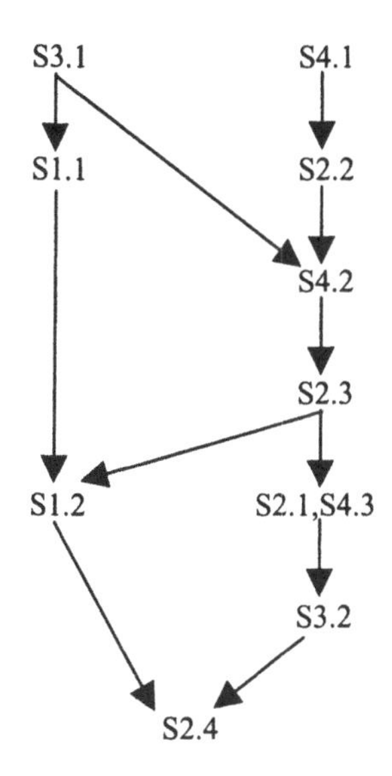

Comment on Baseline Results

The analysis of the 11 strategic options utilising the weightings from the FAE isolates 5 favoured scenarios – S1.1, S2.2, S3.1, S4.1 and S4.2. However, the negotiating group decided to withdraw S1.1 from the evaluation for two reasons:

- its high treatment costs, and
- its unstable ranking.

Its relatively high position in the final pre-order is maninly due to its incomparability with many of the other high ranking options rather than its ability to outrank any of them. This is demonstrated by the discrepancy in its ranking from the two distillations – first in the upward and ninth in the downward. It can only be ranked above two other options – S1.2 and S2.4. This instability occurs to varying degrees with four of the other five sets of weights. It was ranked above the same two options only for the Glarus and Graubunden weightings. For the Ticino and Zurich weightings, it ranked above the same two plus S3.1, S3.2 and S4.3, with St Gallen almost identical to this, with the slight difference that S4.3 was ranked equal to rather than below it.

This leaves us with 4 leading scenarios – S2.2, S3.1, S4.1, S4.2. Their relative positions for the five other weighting groups plus the average weights can be summarised as follows:

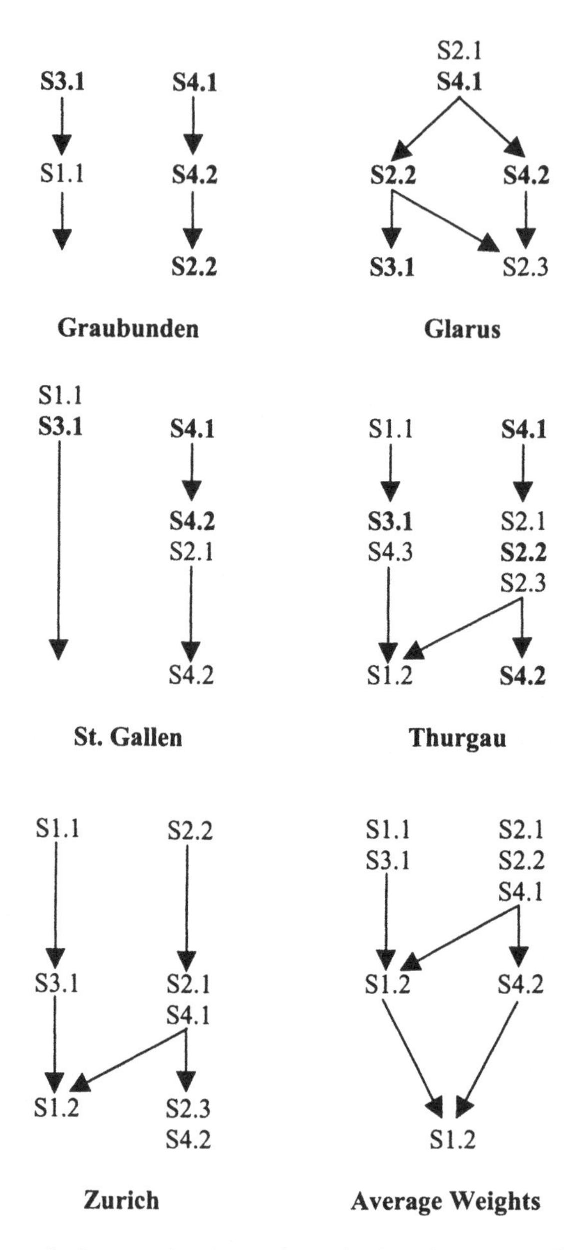

The following conclusion can be drawn from the baseline results for the weighting systems for the six different actors:

S4.1 – This option is ranked first for five of the six weighting groups. The only case where it is in the second rank is in the case of the Zurich weighting set, which emphasises the importance of criteria C1.2 (Energy Use), C2.1 (Cost of Treatment) and C4.1 (Opposition to WTE Plants), where S2.2 performs particularly well. Its outranking performance can thus be said to have a high stability.

S3.1 – Is ranked first three times, in the second rank twice and in the third rank once. However, its high ranking is largely due to its incomparability with many of the better performing options. Taking the results from all weighting sets, S3.1 ranks above only two of the other better performing options – S2.2 and S4.2, and this occurs for only one set. Thus, the stability of its outranking performance is only moderately good.

S2.2 – With the exception of the Zurich weighting set, where this option is in the first rank, S2.2 is consistently placed in the second rank, below S4.1. In only one instance is it placed in the third rank. It consistenly outranks 6 to 8 of the other options over all the weighting sets. It is thus quite stable in its outranking performance.

S4.2 – This option is placed in the second rank twice and in the third rank four times. In all instances it is ranked directly below S4.1, and for four of the six rankings, it is ranked below S2.2. In the majority of weighting systems, it ranks above 5 to 7 of the other options. It is thus quite and stable in its outranking performance

These relative positions are reflected in the ranking of the main options for the average weighting system, with the exception that S2.2 and S4.1 are ranked equal.

6.10 Sensitivity Analysis

For the purposes of a general sensitivity analysis, the effects of certain changes on the results from the baseline case with average weights were assessed. The sensitivity analysis can be broken down into two sections:

- The introduction of veto thresholds
- Variation of the importance weightings of different categories of criteria

Introduction of Veto Thresholds

In the original baseline runs for the six weighting categories and the average weightings, the veto thresholds were muted. As can be seen from Table 6.10, veto thresholds exist for three of the criteria – Transportation Distance (C1.1), Average Treatment Costs (C2.1) and Overcapacity (C3.3).

Introduction of the veto for the three criteria results has an overall benefit in particular for option S4.1, and to a lesser extent S3.1. Both perform well on criterion C1.1. The veto has, however, an unfavourable effect on options S1.1, S1.2 and S2.1 to S2.4, because of their relatively poor performance on C1.1. S1.1 is further penalised because of its veto inducing levels of overcapacity. The overall effect of introducing the veto on the 'average weights' ranking is the relegation of options S3.1 and S2.2 to the second rank. S1.1 drops from the first to the third rank. Both of these are ranked directly below S4.1. The preference in favour of S4.1 is thus emphasised. The benefit to S4.1 over the other favoured options is repeated when the veto is introduced to the analysis of the results for the individual weighting systems from the six actors involved in the negotiation.

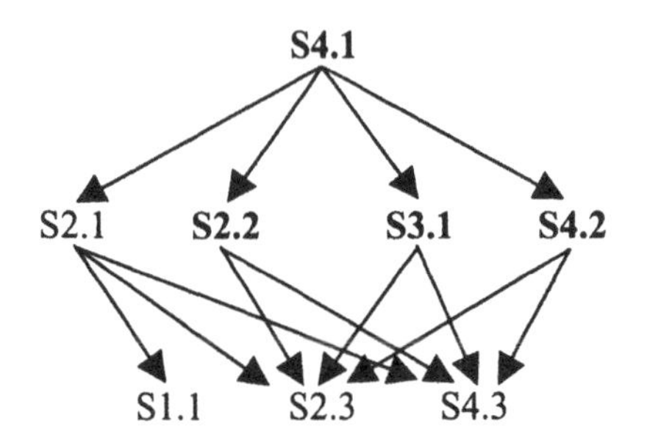

Average Weights
(Veto Thresholds Active)

Varying the Importance of Different Categories of Criteria

This exercise consisted of taking the average weightings for the four categories of criteria – Environmental, Economic, Technical and Political, and observing what effect emphasising the importance of one of the grouping's weights has on the overall ranking. The following four sensitivity tests were carried out:

Test 1 – Increased Importance of Environmental Criteria
The three environmental criteria C1.1, C1.2 and C1.3 were set at their upper quartile values, with the other criteria set at their lower quartile estimates.
Test 2 – Increased Importance of Economic Criteria
The two economic criteria C2.1 and C2.2 were set at their upper quartile values, with the others set at their lower quartile estimates.

Test 3 – Increased Importance of Technical Criteria
The three technical criteria C3.1, C3.2 and C3.3 were set at their upper quartile values, with the others set at their lower quartile estimates.

Test 4 – Increased Importance of Political Criteria
The three political criteria C4.1, C4.2 and C4.3 were set at their upper quartile values, with the others set at their lower quartile estimates.

The weighting systems used in the four tests are detailed below in Table:

Criterion	Test 1	Test 2	Test 3	Test 4	Average Weights
C.1.1	**18**	13	12	13	13
C.1.2	**9**	5	4	4	6
C.1.3	**8**	3	3	3	5
C.2.1	14	**23**	13	14	15
C.2.2	10	**14**	10	10	10
C.3.1	8	8	**16**	7	10
C.3.2	7	8	**15**	7	10
C.3.3	9	9	**12**	8	10
C.4.1	5	5	5	**13**	8
C.4.2	3	3	2	**10**	5
C.4.3	9	9	8	**11**	8
Total	100	100	100	100	100

Table 6 -12 - Weights for Sensitivity Testing

In the ranking for the original average weighting system, S2.2, S3.1 and S4.1 are positioned with S2.1 in the first rank, with S4.2 in the second rank.

In Test 1, S3.1, mainly due to its excellent performance on the Transportation criterion C1.1, ranks above all other options, with S4.1 in the second rank. In general, emphasis on C1.1 penalises options S2.1, S2.2, S2.3 and S2.4. All have prohibitively high scores on this criterion.

In Test 2, S2.2 and S3.1 retain their place in the first rank, with the relatively costly S4.1 relegated to rank 2. S1.1 and S1.2 are the two costliest options, and perform particularly badly criterion C2.1.

In Test 3, S4.1 alone retains its first rank, with S2.2 and S3.1 each performing relatively poorly on one of the technical criteria C3.1 to C3.3.

Both S3.1 and S4.1 are again together in the first rank in Test 4 as a result of their good performance on the political criteria C4.1 to C4.3.

The results of the sensitivity testing can be summarised as follows:

		TEST 1	TEST 2	TEST 3	TEST 4
OPTIONS	S4.1	RANK 2	RANK 2	RANK 1	RANK 1
	S3.1	RANK 1	RANK 1	RANK 2	RANK 1
	S2.2	RANK 3	RANK 1	RANK 2	RANK 2
	S4.2	RANK 4	RANK 3	RANK 3	RANK 3

Table 6 - 13 – Sensitivity Testing Results

This analysis reinforces the stability and consistency associated with the performance of the top two options S3.1 and S4.1.

6.11 Robustness Analysis

By way of contrast with the sensitivity analysis, Table 6-14 below illustrates the robustness of the ranking achieved by the four top options in the basic analysis using the average weighting system. The robustness analysis indicates the range of weightings that do not result in a modification / diminution of the ranking position of the option in question.

The results are as follows:

	Average Weight (%)	S2.2 (%)	S3.1 (%)	S4.1 (%)	S4.2 (%)
CRITERIA					
C1.1	13	7 - 18	12 - 14	7 - 25	10 - 18
C1.2	6	4 - 21	1 - 7	1 - 21	4 - 15
C1.3	5	0 - 10	0 - 12	0 - 12	0 - 40
C2.1	15	13 - 28	15 - 16	0 - 44	13 - 44
C2.2	10	4 - 21	7 - 10	0 - 26	7 - 30
C3.1	10	1 - 12	7 - 25	1 - 25	1 - 13
C3.2	10	8 - 15	0 - 10	0 - 31	0 - 15
C3.3	10	0 - 16	10 - 12	0 - 35	0 - 15
C4.1	8	6 -13	8 - 10	0 - 25	0 - 13
C4.2	5	0 - 8	4 - 20	2 - 20	2 - 8
C4.3	8	0 - 11	7 - 23	0 - 23	0 - 11
Basic Rank		Rank 1	Rank 1	Rank 1	Rank 2
Robustness		Fair	Acceptable	Very Good	Good
Average band width (%)		12	8	25	17

Table 6 -14 – Robustness Analysis

Table 6-14 indicates, for a given criterion, the range of weights over which an option maintains its best rank. The greater this range, the more robust the ranking. The last set of figures on the Table estimates the average of this 'band width' over all the criteria for a given option. It is a direct indicator of the robustness of that option's ranking.

6.12 Overall Comment

S4.1 can be adjudged the best option. It has the top ranking for five out of the six main actors, is consistently in the first or second rank within the sensitivity analysis, and emerges from the robustness analysis as the option which maintains its top ranking over the widest range of criterion weightings.

S3.1 is the second best option, ranking first for three of the main actors, and performs well within the sensitivity analysis. The robustness of its ranking is acceptable, but indicates a relatively narrow range of criterion weighting values over which it will maintain its high ranking. It can be seen that in the case of the weighting system for Canton Glarus, two of the most important weights, C1.3 and C4.1 are outside the ranges in Table 6-14 that are necessary for S3.1 to maintain its optimum ranking. This resulted in S3.1 being relegated to the third rank in this instance.

Option S2.2 completes the shortlist of three, and is the third best option. Except for its top placing with one of the actors, it is otherwise consistently positioned in the second rank. It performs consistently in the sensitivity analysis, only droping to the third rank in one instance. While not as robust as S4.1, it is more so than S3.1, maintaining its optimum rank over an average weighting range of 12%.

It was the decision of the negotiating group that S4.1 be recognised as the preferable option. To cover the situation where this option may be withdrawn for reasons at present unknown to the negotiating group, S3.1 was also submitted to the Swiss Federal Government as a recommended supplementary option.

6.13 References

Banville, C., Landry, M.,Martel, J.M., Boulaire, C. (1993) 'A Stakeholder's Approach to MCDA', University Laval, CRAEDO, *Working paper* 93-77.

Guigou, J.L. (1971) 'On French Location Models for Production Units'. *Regional and Urban Economics*, Volume 1, No. 2, pp107-138.

Maystre, L., Pictet, J., and Simos, J. (1994) *Methodes Multicriteres ELECTRE*. Presses Polytechniques et Universitaires Romandes, Lausanne.

Nijkamp, P. (1975) 'A Multicriteria Analysis for project Evaluation: Economic - Ecological Evaluation of a Land Reclamation Project'. *Papers of the Regional Science Association*, Volume 35, pp87-111.

Rogers, M.G., Bruen, M.P. (1996) 'Using ELECTRE to Rank Options within an Environmental Appraisal - Two Case Studies'. *Civil Engineering Systems*, Vol. 13, pp203-221.

Rogers and Bruen, (1997) 'Threshold values for noise air and water impact criteria within the ELECTRE model'. *Proceedings of the 46th Meeting of the European Working Group on Mullticriteria Aid for Decisions*, Bastia, Corsica, 23rd to 24th October, 1997, Session 6.

Rogers and Bruen, (1998) 'Choosing realistic values of indifference, preference and veto thresholds for use with environmental criteria within ELECTRE'. *European Journal of Operational Research*, Vol. 107, pp542-551.

Roy, B. (1968) 'Classement et choix en presence de points de vue multiples (la methode ELECTRE)'. *Revue Francaise d'Automatique Information et Recherche Operationelle* (RIRO). Volume 8, pp57-75.

Roy, B., Hugonnard, B. (1982) 'Ranking of Suburban Line Extension Projects on the Paris Metro System by a Multicriteria Method' *Transportation Research Record*, Volume 16A, No. 4, pp301-312.

Siskos, J., Hubert, P. (1988) 'Multicriteria Analysis of the Impacts of Energy Alternatives: A Survey and a New Approach'. *European Journal of Operational Research*, Volume 13, pp278-299.

Van Delft, A., Nijkamp, P. (1977) *Multicriteria Analysis and Regional Decision-Making*. Martinus Nijhoff Social Services Division, Leiden.

Vincke, P. (1992) *Multicriteria Decision Aid*. John Wiley, Chichester, United Kingdom.

7 CASE STUDY 3
Selecting the Optimum Route for the Dublin Port Motorway

7.1 Introduction

The Dublin Port Access and Eastern Relief Route (PAERR) was proposed by Dublin Corporation, the local authority with responsibility for road planning in the central city area, as an access route to Dublin Port, which at the same time would provide traffic relief to both the City Centre and nearby residential areas. The route would have to be not only environmentally acceptable but also a sound economic investment. The proposed route is approximately 10 km long, extending from Whitehall on the north side of the city via the Port Area to Booterstown on the south side of the city (Figure 7.1). It was proposed that both ends of the PAERR motorway will connect into a ring road around Dublin which is at present partially completed. Each option considered includes a major crossing of Dublin's main waterway - the River Liffey.

For design and evaluation purposes, the consulting engineers for the project divided the proposed road into three sections (Arup, 1993):

- Northern Section
- Port Section
- Strand (Southern) Section

Various alternative routes were identified for each section:-

- Northern Section : Routes A1 - A6 (6 options)
- Port Section : Routes B1 - B8 (8 options)
- Strand Section : Routes C1 - C3 (3 options)

Each route option in any section could be combined with any route option in an adjacent section. This case study concentrates on choosing the best options for the Port Section of the route. This section involved consideration of a wide range of diverse options differing widely in both environmental and economic terms. No information was available regarding the relative importance of the full set of criteria. ELECTRE IV was therefore chosen as the decision model most suited to the evaluation.

7.2 Route Options

The main project options for the Port Section, B1 to B8, were generated by the consulting engineers in consultation with the client, Dublin Corporation are illustrated in Figure 7.2. They were chosen on the basis of multi-disciplinary workshops comprising engineers, transportation planners, ecologists, landscape architects and environmental planners. B1, B2 and B3 are similar routes, each commencing in the north-west corner of the Port and involving a low level bridge crossing of the Liffey. B4 and B6 also start in the northwest corner of the Port but traverse the river by bored tunnel. B5 and B7 are the most easterly of the options, each involving three alternative river crossing proposals:

- High Navigation Bridge (hb)
- Immersed Tube Tunnel (it)
- Bored Tunnel (bt)

B8 is similar to B1 - B3 but crosses the Liffey further east, with a high shipping clearance bridge spanning diagonally over both the main river and the adjoining river basin.

Route B1 was rejected at an early stage because constructing it would involve the demolition of a school and several houses, the loss of a significant park land, and would cause major severance in one of the large residential areas nearby. Also, for the purposes of the evaluation routes B5 and B7 have been combined because of their close similarities. The three river crossing options for this combined route will still be considered separately. Thus, the eight options evaluated using ELECTRE IV are as follows: B2, B3, B4, B5/7(hb), B5/7(it), B5/7(bt), B6, B8.

7.3 Evaluation Criteria

The above eight project options are assessed within ELECTRE IV on the basis of the following 8 criteria (C_1 to C_8) (7 environmental + cost). These mainly environmental aspects of relevance to the proposed project were identified by the consulting engineers on the basis of the UK Manual of Environmental Appraisal (Department of Transport, 1982). They were defined as follows:

Air Pollution (Air)
Within the assessment, Carbon Monoxide was chosen for the modeling exercise as a general indicator of the effect of each option on air pollution. Values of the Maximum yearly 1 hour average CO Levels were estimated using a simple graphical model (Waterfield and Hickman, 1982).

Noise (Nse)
The areas potentially affected by noise increases due to changes in traffic volumes from the new road were identified, and future noise level increases at these locations were predicted using the UK Department of Transport Noise Prediction Model (Department of Transport, 1983). Noise increases were estimated in decibels for each key location.

Land Use (LU)
Qualitative forecasting of the effects of the new route on development / land use potential was undertaken on the basis of expert judgement by the consulting engineers in conjunction with the Dublin Corporation Planning Department.

The following five point assessment scale was used both for this impact and the four other qualitative ones listed immediately below:

none → minor → moderate → major → severe

On any given impact, the grading assigned for each option by the expert group combines an objective determination of its environmental effect, and a subjective appraisal based on their judgement of what people's perception of the effect will be.

Severance (Sev)
Defined as the level of impediment to existing preferred patterns of movement, the degree of physical severance was assessed qualitatively by the consulting engineers on the basis of estimates by them of the extent of any change to pedestrian journey times caused by an extended journey or increased traffic flow.

Recreation (Rec)
Prediction of the effects of each option on recreational and amenity interests was assessed qualitatively by the consulting engineers on the basis of:

- estimates of the land area to be taken, and
- subjective judgement as to whether this would affect the integrity or value of the recreational interests.

Landscape (Lsc)
The degree of visual intrusion and loss of landscape amenity arising from each option was assessed qualitatively on the basis of:

- examination by the consulting engineers of the engineering design of each proposal

- evaluation by them of site conditions and character
- study by them of similar structures elsewhere, and
- expert judgement by planners / landscape architects.

Construction Disturbance (CnD)
The level of adverse constructional effects on property due to each option was assessed qualitatively on the basis of expert judgement of experienced civil engineering contractors.

Cost (Cst)
Estimates of the construction costs for each option in IR£ were produced by the consulting engineers. Given the preliminary nature of the environmental appraisal, such costs were provisional, as a basis for examining the relative cost implications for the options involved.

7.4 Criterion Valuations

Cost
Options B2 and B3, the two routes including a low-level bridge crossing, were each costed provisionally by the consulting engineers at £70 million. The other routes, all of which pass through the operational areas of the port, were costed at between IR£100m and IR£200m. Of these, the other non-tunnel routes (B5/B7(hb) and B8) were costed by them at in excess of £100 million. All tunneled routes were costed at +£140 million (Arup, 1993). These costs can be summarised as follows:

Options	C_1 - **Cost** (IR million)
B2,B3	70
B5(hb), B7(hb), B8	100+
B4, B5(it), B5(bt), B6, B7(it), B7(bt)	140+

Table 7-1- Cost of Options

Noise Pollution
Only options B2, B3 and B8 were routed through residential areas. As a result, they alone caused significant increases in perceived noise levels. The noise effects of the other options were either minor or non existent. For those options involved, the maximum noise levels estimated at major residential locations in close proximity to the proposed route, based on information from the Ove Arup report (Arup, 1993) are as follows:

Option	2,3	4	6	5it,7it	5bt,7bt	5hb,7hb	8
C_2 - Noise (dB(A))	8	0	0	0	0	2	8

Table 7-2 - Noise Level Increases from Options

Air Pollution Levels

Minor increases in Carbon Monoxide levels were predicted by the consulting engineers for each option (Arup, 1993). Taking account of the information provided by Arup, along with data on typical Carbon Monoxide levels found both on busy open sections of urban roads and in the vicinity of portals of heavily trafficked road tunnels (Hickman and Colwill (1982)), where the Maximum hourly average CO levels for the 'open' routes were approximately 15ppm while the values for the tunneled routes, taken near the entry / exit portal, were estimated at 25ppm (values for the tunneled routes were taken at the tunnel entrance points where concentrations tend to be at their highest), the following estimates of CO levels for each option is as follows:

Option	2,3	4	6	5it,7it	5bt,7bt	5hb,7hb	8
C_3 - Carbon Monoxide (ppm)	15	25	25	25	25	15	15

Table 7-3 - Air Pollution Level Increases from Options

Qualitative impacts

Each of the five qualitative impacts listed above are evaluated on the same five grade points (neutral, minor, moderate, major, severe) against the 'existing situation'. Since, of these, minor, moderate and major can be either beneficial or adverse relative to the existing situation, with the severe grade existing only on the adverse side, the scale in its most detailed form can be expanded to eight grade points as illustrated in the following diagram:

major beneficial	moderate beneficial	minor beneficial	Neutral	minor adverse	moderate adverse	major adverse	severe adverse
MjB	*MdB*	*MnB*	*N*	*MnA*	*MdA*	*MjA*	*SvA*

Scale for Qualitative Impacts

This scale utilises the information directly provided by the consulting engineers within the environmental appraisal. Each option is graded relative to a 'baseline' case, in this case the 'do-nothing' situation.

Using this scale form, the consulting engineers, in partnership with the relevant expert groups, produced the following grades for each of the five qualitatively assessed impacts:

		Option						
		2,3	4	6	5it,7it	5bt,7bt	5hb,7hb	8
Criteria	C_4.LU	*MdA*	*N*	*MnA*	*MnA*	*MnA*	*MnA*	*MnA*
	C_5.Sev	*MnB*	*MnB*	*MnB*	*MnB*	*MnB*	*MnB*	*MnA*
	C_6.Rec	*MdA*	*N*	*N*	*MnA*	*MnA*	*MnA*	*N*
	C_7.Lsc	*MnA*	*N*	*N*	*N*	*N*	*MjA*	*SvA*
	C_8.CnD	*MdA*	*MnA*	*MnA*	*MjA*	*MnA*	*MdA*	*MdA*

Table 7-4 - Ratings of each option for Qualitatively Assessed Environmental Impacts

7.5 Threshold Estimates

In order to use the ELECTRE IV decision model, estimates of the indifference, preference and veto thresholds are required for each criterion. These are estimated for the three quantitative and five qualitative criteria as follows:

(i) Quantitatively Evaluated Criteria
For all three quantitative criteria, it should be noted that preference is expressed in favour of the option within any given pair with the lower valuation.

Cost
Hokkanen and Salminen (1997) suggested that an error of 10% for total cost estimates was quite usual. Given the preliminary nature of this assessment, we have used 15%, and placed the indifference threshold at this level. The preference threshold has been put at twice this value - 30%. The veto threshold has been placed at 90% of the higher estimate. Thus:

Indifference threshold q(g(b))	0.15 g(b)
Preference threshold p(g(b))	0.30 g(b)
Veto threshold v(g(b))	0.90 g(b)

Table 7-5 - Thresholds for Cost criterion

Noise
Rogers and Bruen (1998) estimated the indifference, preference and veto thresholds for the noise criterion on the 18 hour A-weighted decibel scale as follows:

Indifference threshold q(g(b))	3dB(A)
Preference threshold p(g(b))	6dB(A))
Veto threshold (g(b))	15dB(A)

Table 7-6 - Thresholds for Noise criterion

Air Pollution (Carbon Monoxide)
All three thresholds have been estimated by Rogers and Bruen (1997) as follows. The indifference threshold for estimated values of Carbon Monoxide has been set at 50%

of the higher of the two valuations being compared. The preference threshold has two requirements, firstly that the two valuations can be distinguished, i.e. the difference between the two be greater than the indifference threshold, and secondly that the higher of the two valuations be greater than the 100 index value - the level that induces a health effect in humans, valued at a Maximum 1-hour average Carbon Monoxide concentration of 35 parts per million. The veto is invoked if the higher value is at the 300 index value - defined as the 'unhealthful' level of CO, set at a Maximum 1-hour average concentration of 120 parts per million (Canter, 1995).

Indifference threshold	**if g(b) - g(a) > 50% g(b)**
Preference threshold	**if g(b) - g(a) > 50% g(b) and g(b) > 35ppm** (35ppm is the 'threshold' 1 hour peak CO level)
Veto threshold (v)	**if g(b) - g(a) > 50% g(b) and g(b) > 120ppm** (120ppm is the 'unhealthful' 1 hour peak CO level)

Table 7-7 - Thresholds for Air Pollution criterion (Carbon Monoxide)

(ii) Qualitatively Evaluated Impacts

If qualitatively assessed environmental impacts are to be used in the ELECTRE IV model alongside quantitative impacts such as air and noise referred to above, a method of determining the indifference preference and veto thresholds for them must be devised which is compatible with that devised for the mathematically modeled impacts by Rogers and Bruen (1997).

The grading system devised by the authors for qualitatively assessing impacts in the PAERR case study requires the information for each option regarding their performance to be placed on the eight-point scale noted above. The grade obtained can then be used as a basis for assessing the preference relationship between it and the existing situation, be it indifference, preference or veto inducing. The method achieves this by matching the impact grades to the three criterion thresholds required by ELECTRE IV.

Thus, a minor impact indicates the grade at which the change from the existing situation is perceptible. It delineates the end of the zone of indifference and the start of the zone of preference. A moderate impact is one whose scale of change is such that clear preference can be expressed relative to the existing situation. If the impact is moderate adverse, the baseline situation is clearly preferable to the one induced by this option, and vice versa for moderate beneficial. The major adverse / beneficial impact is the first grade within the zone of veto. Relative to the baseline situation, the preference inferred becomes so strong it begins to override positive results on other impacts for the badly performing option. At the 'severe adverse' grade, the environmental effect is so adverse that the option becomes untenable, regardless of its performance on the other impacts within the analysis. It thus defines the extreme point of the veto zone.

For a given impact, each expert assigns each option, in comparison to the existing situation, to one of the eight available grades. An analysis of the data enables a mean and standard error score for each option to be compiled. The results of this analysis are then used to assess the direct preference relationship for every possible pairing of options. The scale can be illustrated graphically as follows:

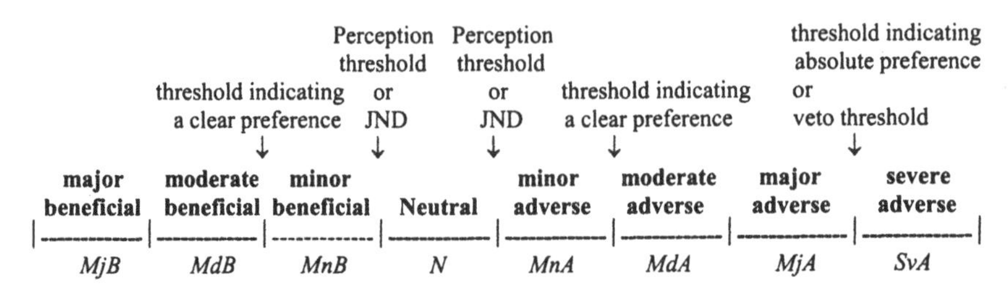

Qualitative Impact Grading Scale

Using these eight grades the preference relationships between any pair of options can be now defined as follows:

Indifference:
All options graded neutral can be deemed to be indifferent to one another, as the difference between them and the 'null option' existing situation is barely perceptible, and is still firmly in the indifference range where no preference can be allocated.

Veto
If the mean score for one of the two options is graded at 'severe adverse', that option is automatically ranked last. ($D_j(a,b) = 1$)

Preference:
Data variability
Before any preference relationship can be confirmed, we must cross check against the variability of the data. In the absence of any measure of the variability in the data from the qualitative assessments on the Port Access Motorway, we assume a standard deviation in the data of one grade. Thus, a difference of one will not be considered convincing evidence of a preference, and we set the indifference threshold at this figure, with weak preference deemed to exist at one additional grade. We thus require a difference equal to at least two grades before a preference relationship can exist. The quality of that preference relationship, be it weak or strong, will then depend on the relative grading of the options as outlined below.

Thus, as with the quantitatively assessed impacts air and water, where the indifference threshold was assessed on the basis of model error, so too in the case of the qualitatively assessed impacts, where this threshold is based on the standard deviation of the responses from the expert groupings.

Weak Preference
Weak preference in favour of the better performing option can exist in the case where one is ranked at '**minor adverse / beneficial**', and the other is a distance away from it equal to the indifference threshold plus one grade, i.e. a minimum of two grades. (C_j (a,b) = 0.5)

Strict Preference
Strict preference in favour of the better performing option can exist in the case where one is ranked at '**moderate adverse / beneficial**', and the other is a distance away from it equal to the indifference threshold plus one grade, i.e. a minimum of two grades. ($C_j(a,b) = 1.0$)

On a given impact, the strict preference threshold thus requires one of the two options being examined to be a minimum of two grades beyond neutral, with a minimum grade difference required between the two options equal to the standard deviation in the survey data plus one grade.

Indifference threshold	A difference of one grade (assuming this to be the standard deviation of the data from the expert survey)
Preference threshold	A difference of one grade more than the standard deviation in the data, with one of the options graded at 'moderate' or further beyond neutral
Veto threshold	Full veto is induced if one option is at neutral or better and the other is ranked at 'severe adverse'

Table 7-8 - Thresholds for Qualitatively Assessed Criteria

A simpler set of thresholds, with indifference set at ± 1 grade and preference at ± 2 grades, using the same general format as for the qualitative criteria in the previous case study, would yield very similar results to those obtained using the values in Table 7-8.

7.6 Criterion Relations

Based on the criterion scores for each of the options in Tables 7.1, 7.2. 7.3 and 7.4, and the criterion thresholds as defined in Tables 7.5, 7.6, 7.7 and 7.8, the following strong and weak preference, indifferent and equal relationships were deduced for each pair of the decision criteria.

It should be noted that, for the qualitative criteria, a standard deviation of one grade in the data was assumed.

Strict Preference Relationships

For each pair of options (a,b), the number of criteria for which a is strictly preferred to b, $m_p(a,b)$, is given in the following matrix:

	2,3	4	6	5it,7it	5bt,7bt	5hb,7hb	8
2,3	—	1	1	1	1	1	1
4	3	—	0	1	0	1	2
6	2	0	—	1	0	1	2
5it,7it	1	0	0	—	0	1	2
5bt,7bt	1	0	0	1	—	1	2
5hb,7hb	0	0	0	0	0	—	0
8	1	0	0	0	0		—

Table 7-9 - Number of Strict Preferences attributable to each pair of Options (a,b)

Weak Preference Relationships

For each pair of options (a,b), the number of criteria for which a is weakly preferred to b, $m_q(a,b)$, is given in the following matrix:

	2,3	4	6	5it,7it	5bt,7bt	5hb,7hb	8
2,3	—	1	1	1	1	1	2
4	0	—	0	0	0	0	1
6	0	0	—	0	0	0	1
5it,7it	0	0	0	—	0	0	1
5bt,7bt	0	0	0	0	—	0	1
5hb,7hb	1	2	2	2	2	—	2
8	0	2	2	2	2	0	—

Table 7-10 - Number of Weak Preferences attributable to each pair of Options (a,b)

Indifferent but Measurably Better Relationships

For each pair of options (a,b), the number of criteria for which a and b are indifferent, $m_i(a,b)$, even though the measured performance of a is better, is given in the following matrix:

	2,3	4	6	5it,7it	5bt,7bt	5hb,7hb	8
2,3	—	0	0	1	0	0	0
4	2	—	1	2	2	4	2
6	3	0	—	1	1	3	1
5it,7it	3	0	0	—	0	1	0
5bt,7bt	4	0	0	0	—	2	1
5hb,7hb	2	0	0	1	0	—	1
8	1	0	0	2	1	1	—

Table 7-11 - Number of Indifferent Relationships attributable to each pair of Options (a,b), with the performance of a measurably better than b.

Equal Pairings

For each pair of options (a,b), the number of criteria on which a and b perform identically, $m_0(a,b)$, is given in the following matrix:

	2,3	4	6	5it,7it	5bt,7bt	5hb,7hb	8
2,3	—	1	1	1	1	3	3
4	0	—	7	5	6	1	1
6	0	0	—	6	7	2	2
5it,7it	0	0	0	—	7	3	1
5bt,7bt	0	0	0	0	—	3	1
5hb,7hb	0	0	0	0	0	—	4
8	0	0	0	0	0	0	—

Table 7-12- Number of identical performances for each pair of Options (a,b)

Sample Relationships

In order to illustrate how the four different types of relationships are derived, sample calculations of strong and weak preference, indifference and equality have been given for two of the criteria, Cost and Noise:

C_1: Cost
Option 2,3 is strictly preferred to Options 4, 6, 5/7bt, 5/7it
For example:
$g_1(2) \quad = IR£70m$
$g_1(4) \quad = IR£140m$
$p_1 g_1(4) \quad = 0.30 * 140 \ = 42$
since $g_1(4) - g_1(2) > 42$, 2P4

Options 5/7hb and 8 are weakly preferred to Options 4, 6, 5/7bt, 5/7it
For example:
$g_1(8)$ = IR£100m
$g_1(4)$ = IR£140m
$p_1g_1(4)$ = 0.30 * 140 = 42
$q_1g_1(4)$ = 0.15 * 140 = 21
since $g_1(4) - g_1(8) = 40$, and $q_1g_1(4) < 40 < p_1g_1(4)$, 8Q4

Option 2,3 is weakly preferred to Options 5/7hb, 8

C_2: Noise
Options 4, 6, 5/7bt, 5/7it are strictly preferred to Options 2, 3 and 8
For example:
$g_2(4)$ = 0 dB(A)
$g_2(2)$ = 8 dB(A)
$p_2g_2(4)$ = 6 dB(A)
since $g_2(2) - g_2(4) = 8 > p_2g_2(4)$, 4P2

Option 5/7hb is weakly preferred to Options 2, 3 and 8
For example:
$g_2(5/7_{hb})$ = 2 dB(A)
$g_2(2)$ = 8 dB(A)
$p_2g_2(2)$ = 6 dB(A)
$q_2g_2(2)$ = 3 dB(A)
since $g_2(2) - g_2(5/7_{hb}) = 6$, and $q_2g_2(2) < 6 \leq p_2g_2(2)$, $5/7_{hb}$Q2

Options 4, 6, 5/7it and 5/7bt is indifferent to, but measurably better than, Option 5/7hb
For example:
$g_2(5/7_{hb})$ = 2 dB(A)
$g_2(4)$ = 0 dB(A)
$g_2(5/7_{hb}) - g_2(4) = 2 < q_2g_2(5/7_{hb})$, 4I$5/7_{hb}$

Options 2, 3 and 8 are equal, as also are Options 4, 6, 5/7it, 5/7bt.

7.7 Outranking Relationships for Baseline Threshold Values

One of the following outranking relationships can be deduced for each pair of options where:

- Quasi-dominance (aS_qb),
- Canonical-dominance (aS_cb),
- Pseudo-dominance (aS_pb),
- Sub-dominance (aS_sb), and

- Veto-dominance (aS_Vb),

Each of these are defined in detail in Chapter 3.

On the basis of the information in Tables 7.9 to 7.12, the following outranking relationships were deduced:

	2,3	4	6	5it,7it	5bt,7bt	5hb,7hb	8
2,3	—					S_c	
4		—	S_q	S_q	S_q	S_s	S_p
6		S_c	—	S_q	S_q	S_s	S_p
5it,7it				—		S_s	S_p
5bt,7bt		S_p	S_c	S_q	—	S_s	S_p
5hb,7hb						—	S_q
8						S_s	—

Table 7-13- Outranking Relationships for each pair of Options (a,b)

As stated in Chapter 3, for every outranking relationship, the degree of credibility of outranking can be estimated as follows:

- if aS_qb, then S(a,b) = 1
- if aS_cb, then S(a,b) = 0.8
- if aS_pb, then S(a,b) = 0.6
- if aS_sb, then S(a,b) = 0.4
- if aS_vb, then S(a,b) = 0.2

with S(a,b) = 0 if none of these above five relationships apply

On the basis of these valuations, the following degree of credibility matrix is complied:

	2,3	4	6	5it,7it	5bt,7bt	5hb,7hb	8
2,3	—	0	0	0	0	0.8	0
4	0	—	1	1	1	0.4	0.6
6	0	0.8	—	1	1	0.4	0.6
5it,7it	0	0	0	—	0	0.4	0.6
5bt,7bt	0	0.6	0.8	1	—	0.4	0.6
5hb,7hb	0	0	0	0	0	—	1
8	0	0	0	0	0	0.4	—

Table 7-14- Degree of Credibility Matrix

The discrimination threshold s(λ) utilised is constant and equal to 0.1. This enables the following credibility differentials to be calculated:

	λ_l	λ_l - s(λ_l)	λ_{l+1}
aS_qb	1.0	0.9	0.8
aS_cb	0.8	0.7	0.6
aS_pb	0,6	0.5	0.4
aS_sb	0.4	0.3	0.2
aS_vb	0.2	0.1	0.0

Table 7-15 - Calculation of the Degrees of Credibility

Downward Distillation

$A_0 = A$; $s(\lambda) = 0.1$

- First distillation

Step 1

$l = 0$; $D_0 = A_0$; $\lambda_0 = 1$, $\lambda_0 - s(\lambda_0) = 0.9$, $\lambda_1 = 0.8$

	2/3	**4**	**6**	**5/7it**	**5/7bt**	**5/7hb**	**8**
S		6, 5/7it, 5/7bt	5/7it, 5/7bt		5/7it	8	
Strength	0	3	2	0	1	1	0
Weakness	0	0	1	3	2	0	1
Qualification	0	3	1	-3	-1	1	-1

$C_1 = D_1 = \{4\}$

$A_1 = A_0 \setminus C_1 = \{2/3, 6, 5/7it, 5/7bt, 5/7hb, 8\}$

- Second distillation

Step 1

$l = 0$; $D_0 = A_1$; $\lambda_0 = 1.0$, $\lambda_0 - s(\lambda_0) = 0.9$, $\lambda_1 = 0.8$

	2/3	**6**	**5/7it**	**5/7bt**	**5/7hb**	**8**
S		5/7it, 5/7bt		5/7it	8	
Strength	0	2	0	1	1	0
Weakness	0	0	2	1	0	1
Qualification	0	2	-2	0	1	-1

$C_2 = D_1 = \{6\}$

$A_2 = A_1 \setminus C_2 = \{2/3, 5/7it, 5/7bt, 5/7hb, 8\}$

- Third distillation

Step 1

$l = 0$; $D_0 = A_2$; $\lambda_0 = 1.0$, $\lambda_0 - s(\lambda_0) = 0.9$, $\lambda_1 = 0.8$

	2/3	5/7it	5/7bt	5/7hb	8
S			5/7it	8	
Strength	0	0	1	1	0
Weakness	0	1	0	0	1
Qualification	0	-1	1	1	-1

$D_1 = \{5/7bt, 5/7hb\}$

Step 2

$l = 1$; $D_1 = \{5/7bt, 5/7hb\}$; $\lambda_1 = 0.4$, $\lambda_0 - s(\lambda_0) = 0.3$, $\lambda_2 = 0.2$

$D_2 = C_3 = \{5/7bt\}$

$A_3 = A_2 \setminus C_3 = \{2/3, 5/7it, 5/7hb, 8\}$

- Fourth distillation

Step 1

$l = 0$; $D_0 = A_3$; $\lambda_0 = 1.0$, $\lambda_0 - s(\lambda_0) = 0.9$, $\lambda_1 = 0.8$

	2/3	5/7it	5/7hb	8
S			8	
Strength	0	0	1	0
Weakness	0	0	0	1
Qualification	0	0	1	-1

$D_1 = C_4 = \{5/7hb\}$

$A_4 = A_3 \setminus C_4 = \{2/3, 5/7it, 8\}$

- Fifth distillation

Step 1

$l = 0$; $D_0 = A_4$; $\lambda_0 = 0.6$, $\lambda_0 - s(\lambda_0) = 0.5$, $\lambda_1 = 0.4$

	2/3	5/7it	8
S		8	
Strength	0	1	0
Weakness	0	0	1
Qualification	0	1	-1

$D_1 = C_5 = \{5/7it\}$

$A_5 = A_4 \setminus C_5 = \{2/3, 8\}$

- Sixth distillation

$l = 0$; $D_1 = A_5$; $\lambda_0 = 0.0$, $\lambda_0 - s(\lambda_0) = 0.0$, $\lambda_2 = 0.0$

$D_1 = C_6 = \{2/3, 8\}$

The last two options cannot be separated

The resultant ranking obtained from the downward distillation procedure is as follows:

Upward Distillation

$A_0 = A$; $s(\lambda) = 0.1$

- First distillation

Step 1
$l = 0$; $D_0 = A_0$; $\lambda_0 = 1$, $\lambda_0 - s(\lambda_0) = 0.9$, $\lambda_1 = 0.8$

	2/3	**4**	**6**	**5/7it**	**5/7bt**	**5/7hb**	**8**
S		6, 5/7it, 5/7bt	5/7it, 5/7bt		5/7it	8	
Strength	0	3	2	0	1	1	0
Weakness	0	0	1	3	2	0	1
Qualification	0	3	1	-3	-1	1	-1

$C_1 = D_1 = \{5/7it\}$
$A_1 = A_0 \setminus C_1 = \{2/3, 4, 6, 5/7bt, 5/7hb, 8\}$

- Second distillation

Step 1
$l = 0$; $D_0 = A_1$; $\lambda_0 = 1.0$, $\lambda_0 - s(\lambda_0) = 0.9$, $\lambda_1 = 0.8$

	2/3	4	6	5/7bt	5/7hb	8
S		6, 5/7bt	5/7bt		8	
Strength	0	2	1	0	1	0
Weakness	0	0	1	2	0	1
Qualification	0	2	0	-2	1	-1

$C_2 = D_1 = \{5/7bt\}$
$A_2 = A_1 \setminus C_2 = \{2/3, 4, 6, 5/7hb, 8\}$

- Third distillation

Step 1
$l = 0$; $D_0 = A_2$; $\lambda_0 = 1.0$, λ_0 - $s(\lambda_0) = 0.9$, $\lambda_1 = 0.8$

	2/3	4	6	5/7hb	8
S		6		8	
Strength	0	1	0	1	0
Weakness	0	0	1	0	1
Qualification	0	1	-1	1	-1

$D_1 = \{6, 8\}$
Step 2
$l = 1$; $D_1 = \{6, 8\}$; $\lambda_1 = 0.6$, λ_0 - $s(\lambda_0) = 0.5$, $\lambda_2 = 0.4$
$D_2 = C_3 = \{8\}$
$A_3 = A_2 \setminus C_3 = \{2/3, 4, 6, 5/7hb\}$

- Fourth distillation

Step 1
$l = 0$; $D_0 = A_3$; $\lambda_0 = 1.0$, λ_0 - $s(\lambda_0) = 0.9$, $\lambda_1 = 0.8$

	2/3	4	6	5/7hb
S		6		0
Strength	0	1	0	0
Weakness	0	0	1	0
Qualification	0	1	-1	0

$D_1 = C_4 = \{6\}$
$A_4 = A_3 \setminus C_4 = \{2/3, 4, 5/7hb\}$

- Fifth distillation

Step 1
$l = 0$; $D_0 = A_4$; $\lambda_0 = 0.8$, λ_0 - $s(\lambda_0) = 0.7$, $\lambda_1 = 0.6$

	2/3	4	5/7hb
S	5/7hb		
Strength	1	0	0
Weakness	0	0	1
Qualification	1	0	-1

$D_1 = C_5 = \{5/7hb\}$
$A_5 = A_4 \setminus C_5 = \{2/3, 4\}$

- Sixth distillation

$l = 0$; $D_1 = A_5$; $\lambda_0 = 0.0$, $\lambda_0 - s(\lambda_0) = 0.0$, $\lambda_2 = 0.0$
$D_1 = C_6 = \{2/3, 4\}$

The last two options cannot be separated

The resultant ranking obtained from the upward distillation procedure is as follows:

Final Ranking

The upward and downward distillations are combined to give the overall ranking for the base case as follows:

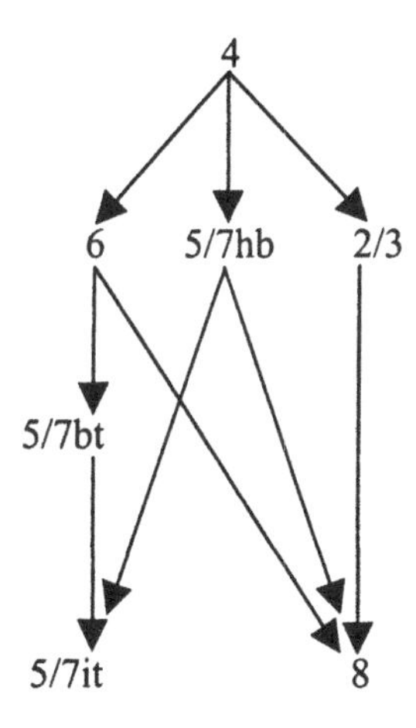

Comment on Baseline Ranking

The above final ranking shows Option 4 ranked first. There is a definitive ranking among the tunneled options, 4 → 6 → 5/7bt → 5/7it, while within the overland options 2/3 and 5/7hb both rank above 8 but cannot be compared. However, there are very few comparisons that can be made between the two groupings, except that Option 4 ranks above all the overland option. Except for the ranking of 6 above 8, all the other tunneled options ranked below 4 cannot be compared directly with any of the land / bridge options. Therefore, choice of second best option is not straightforward, as three options exist, 2/3, 6 and 5/7hb, are all ranked directly below 4, but cannot be separated in the ranking procedure.

7.8 Sensitivity Testing on Threshold Values

Introduction

In an effort to try and separate the three second ranked options, a series of sensitivity test were carried out in which the baseline thresholds within the original analysis were varied. These tests can be summarised in terms of the five categories of tests shown below in Table 7-16:

	TEST 1		TEST 2		TEST 3		TEST 4	
Criterion	**Q**	**p**	**q**	**p**	**q**	**p**	**q**	**p**
Cost (IR£m)	Base	base	.1g(a)	.15g(a)	base	base	.3g(a)	.6g(a)
Noise (dB(A))	0	3	base	base	6	9	base	base
Air (ppm)	.25g(a)	20	base	base	.5g(a)	50	base	base
------------------	------------------		------------------		------------------		------------------	
Land Use								
Severance								
Recreation	s. of data = 0		base		s.d. of data = 2		base	
Landscape								
Construction								
Disturbance								

Table 7-16 - Thresholds for Sensitivity Testing

Tests 1 involved leaving the thresholds for the cost criterion unchanged and lowering the thresholds for the other seven environmental / technical criteria, thus making the model more sensitive to score differences on the environmental / technical criteria. Test 2 lowered the thresholds for the cost criteria while leaving the others unaltered. This had the opposite affect on increasing the sensitivity of the model to the relative cost of the options. Test 3 kept the cost thresholds at their original levels while increasing them for the other environmental / technical criteria, thus making thr model less sensitive to differences in their scores. Finally, Test 4 increased the threshold values for the cost criterion while leaving the others at their original values, thus lessening the impact on the model of cost differences.

Outranking Relationships for the Sensitivity Tests

The Outranking Relationships for each of the for tests outlined above were as follows:

	2,3	**4**	**6**	**5it,7it**	**5bt,7bt**	**5hb,7hb**	**8**
2,3	—						
4		—	Sq	Sq	Sq		
6			—	Sq	Sq		
5it,7it				—			
5bt,7bt		Ss	Ss	Sq	—		
5hb,7hb						—	Sc
8							—

Table 7-17 - Sensitivity Test 1

	2,3	4	6	5it,7it	5bt,7bt	5hb,7hb	8
2,3	—					Sp	
4		—	Sq	Sq	Sq		
6		Sc	—	Sq	Sq		
5it,7it				—			
5bt,7bt		Sp	Sc	Sq	—		
5hb,7hb						—	Sc
8						Ss	—

Table 7-18 - Sensitivity Test 2

	2,3	4	6	5it,7it	5bt,7bt	5hb,7hb	8
2,3	—	Sp	Sp		Sp	Sc	Sq
4		—	Sq	Sq	Sq	Sp	Sc
6		Sc	—	Sq	Sq	Sp	Sc
5it,7it		Sp	Sp	—	Sc	Sp	Sc
5bt,7bt		Sp	Sc	Sq	—	Sp	Sc
5hb,7hb						—	Sq
8							—

Table 7-19 - Sensitivity Test 3

	2,3	4	6	5it,7it	5bt,7bt	5hb,7hb	8
2,3	—					Sp	
4	Sp	—	Sq	Sq	Sq	Sp	Sc
6	Sp	Sc	—	Sq	Sq	Sp	Sc
5it,7it	Ss			—		Sp	Sc
5bt,7bt	Ss	Sp	Sc	Sq	—	Sp	Sc
5hb,7hb						—	Sq
8						Sp	—

Table 7-20 - Sensitivity Test 4

Results of Sensitivity Testing

The ranking from Test 1 is illustrated below:

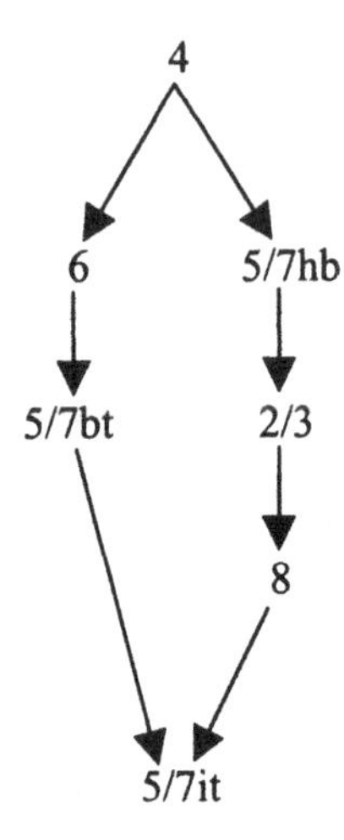

The emphasis on score differences for the environmental / technical criteria results in the model separating the high-bridge option 5/7hb from the land / low bridge option 2/3, with 5/7hb scoring better on the non-economic criteria. Option 4 remains ranked first.

The ranking from Test 2 is illustrated below:

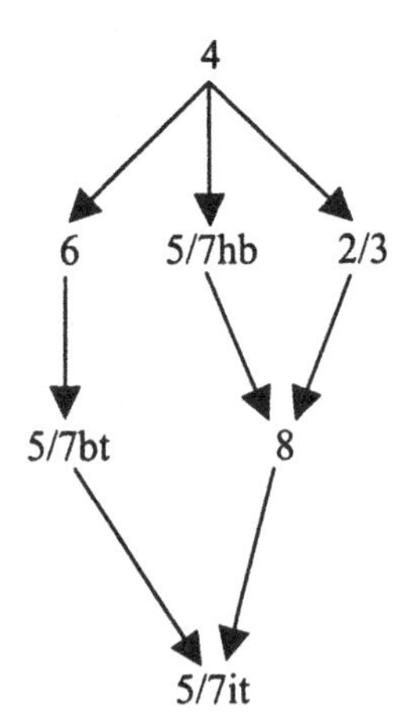

Here, the emphasis on cost differences has resulted in 2/3, one of the lowest cost options, returning to the second rank alongside Options 6 and 5/7hb. All three are incomparable to each other. Option 4 remains ranked first.

The ranking from Test 3 is illustrated below:

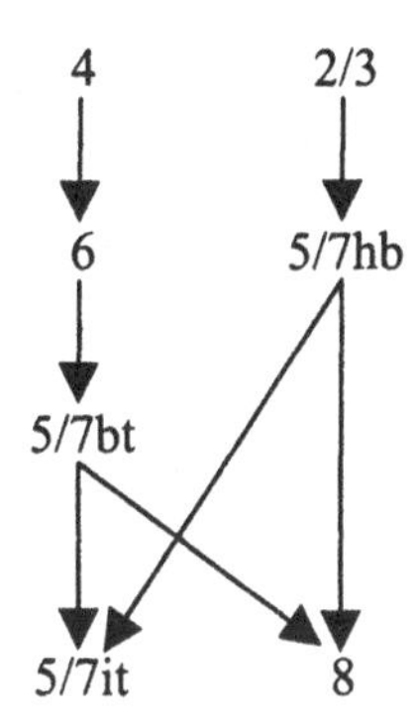

In this case, the de-emphasising of differences on the environmental / technical criteria has resulted in Option 2/3, one of the lowest cost options, no longer being ranked below Option 4 and ranking above the more expensive land / bridge based Option 5/7hb.

Test 4 yields the same ranking as the baseline case.

7.9 Overall Conclusions

The core evaluation indicates clearly that 4 is the best ranked option. This is borne out in the sensitivity testing, which shows no instance where any other option is ranked above it.

Only when the differences in the cost criterion scores are emphasised does 2/3 join it in the first rank.

The evaluation identifies clearly that Options 5/7bt, 5/7it and 8 are the worst performing options.

The model indicates the low level of comparison possible between options 2/3, 6 and 5/7hb. In the baseline case, all are in the second rank. As the cost thresholds are lowered or the environmental / technical thresholds increased, 2/3 gradually moves above 5/7hb, with the opposite occurring as the environmental / technical thresholds are decreased or the cost thresholds are increased. At no stage is a direct ranking between the tunnel options 6 and 5/7bt and the overland / bridge options 2/3 and 5/7hb established.

7.10 References

Arup and Partners (1993) Port Access and Eastern Relief Route: Environmental Appraisal Report. Dublin Corporation.

Canter, L. (1995) *Environmental Impact Assessment*. Mc Graw Hill, New York, 2nd Edition.

Department of Transport (1982) *Manual of Environmental Appraisal*. HMSO, London

Department of the Transport (1983) *Calculation of Road Traffic Noise*. HMSO, London.

Hickman, A.J. and Colwill, D.M. (1982) 'The estimation of air pollution from road traffic'. *Transport and Road Research Laboratory Report, LR1052*, Crowthorne. UK

Hokkanen, J. and Salminen, P. (1997) *Choosing a solid waste management system using multicriteria decision analysis'*. European Journal of Operational Research, Vol. 98, p19-36, Elsevier B.V.

Rogers M.G., and Bruen, M.P. (1997) 'Threshold values for noise air and water impact criteria within the ELECTRE model'. *Proceedings of the 46th Meeting of the European Working Group on Mullticriteria Aid for Decisions*, Bastia, Corsica, 23rd /f 24th October, 1997, Session 6.

Rogers, M.G., and Bruen, M..P. (1998) 'Choosing Realistic Values of Indifference, Preference and Veto Thresholds for Use With Environmental Criteria within ELECTRE', *European Journal of Operational Research*, Vol. 107, No. 3, pp. 542-551.

Waterfield, V.H. and Hickman, A.J. (1982) *Estimating Air Pollution from Road Traffic: A Graphical Screening Method*. Transport and Road Research Laboratory Supplementary Report No. 752, Crowthorne, Berkshire.

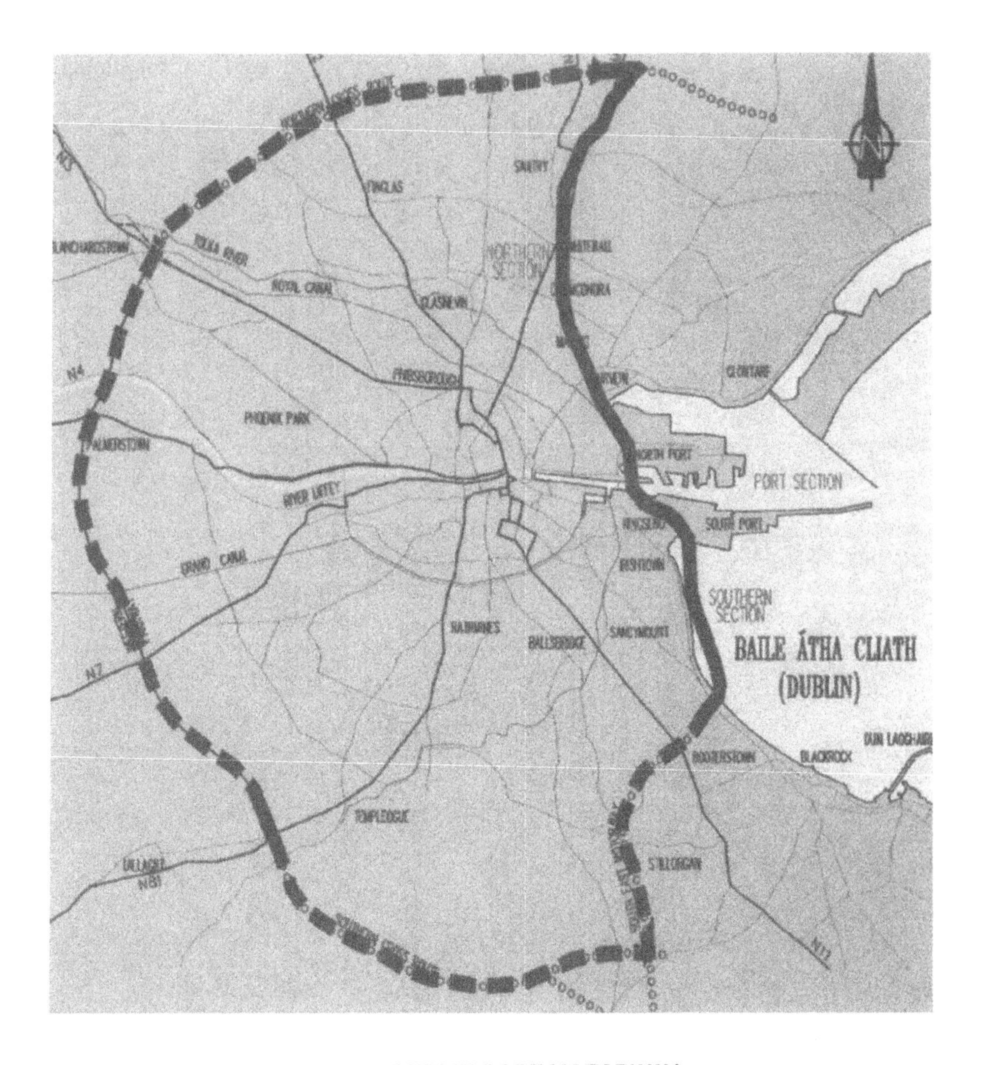

DUBLIN PORT MOTORWAY

DUBLIN RING ROAD

Fig.7.1

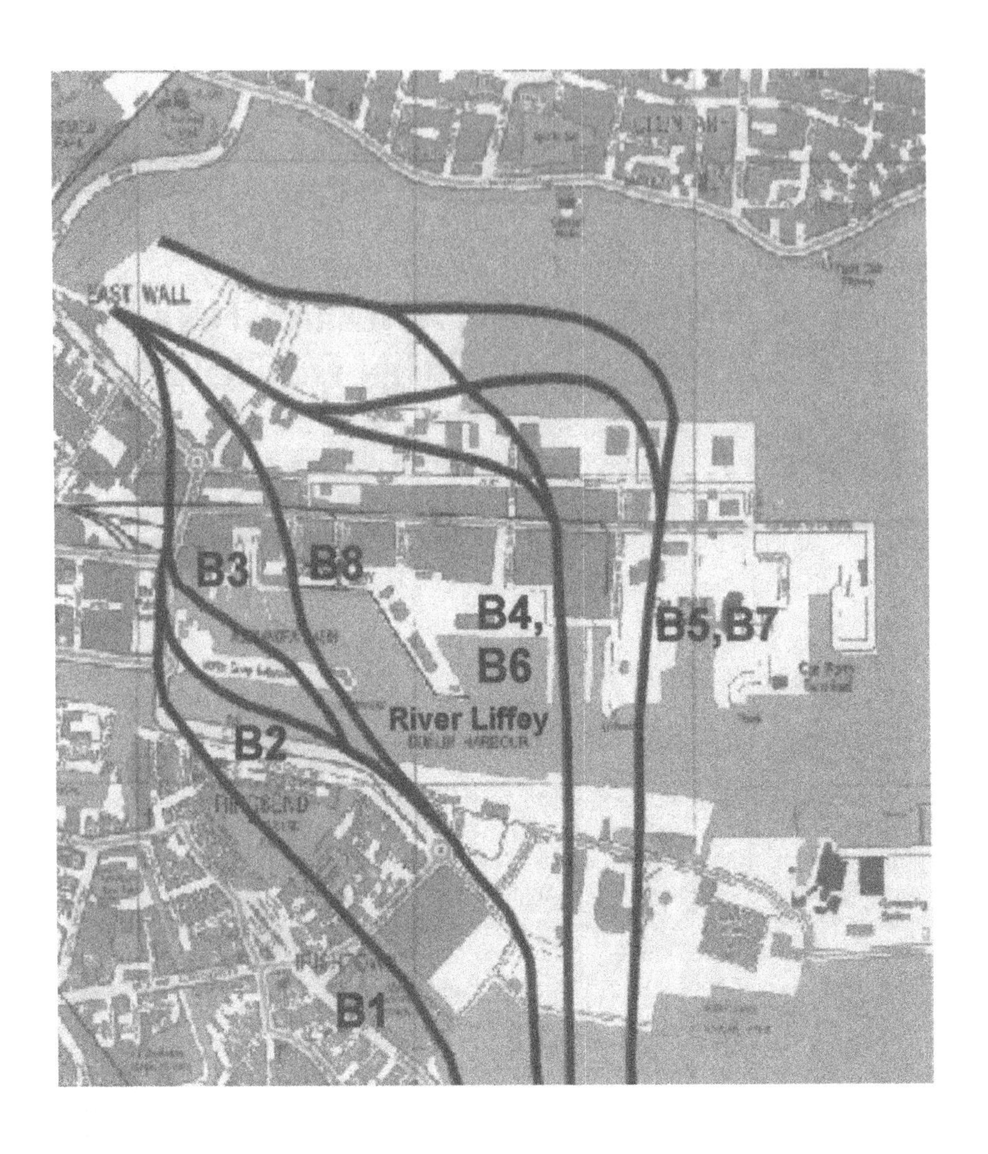

Route Options

Fig.7.2

8 NEW IDEAS AND TECHNIQUES WITHIN DECISION AID

8.1 Models and Decision Aid

A decision model strives to represent reality in a rational, logical, coherent and purposeful way. Its primary purpose, in the context of civil and environmental engineering, is to predict the consequences of proposed infrastructure projects. The output from the model can then be used to influence the opinions and decisions of the relevant actors. It is a tool used by the decision-makers to help achieve common set of agreed goals and objectives. It can thus be termed a construct. The model (or construct) in question may be the property of an individual, a group of individuals, a wider community such as the inhabitants of a major city or the general scientific community.

The outcomes that result from the coherent and logical treatment of the data by the model are valid only for the person or group of persons who chose the model for that particular purpose. A different group may select a different type of model, may use a different version of the same type of model, or may take the same results and interpret them in a different way.

Given that the end product of using a model such as ELECTRE is a decision and recommendations, use of the model can only be an aid to decision making, since it is the actors involved in the process who actually make the decision. The 'decision' does not exist as an entity in itself, and does not involve the 'unveiling' or discovery of some hidden truth. Rather it is the result of consultations taking place between the actors, with the model helping to make sense of the project information at their disposal.

Given that models can aid in the making of better decisions in the presence of ambiguity and uncertainty, what makes ELECTRE particularly applicable to the type of problem developed in this book?

The ELECTRE decision model can help ensure that the decision is framed in a correct way for the group of decision-makers or negotiators:

- ELECTRE is most appropriately used when a relatively large number of competing options need to be short-listed to a smaller number of preferred ones (Rogers and Bruen, 1995). It thus requires a minimum number of options to be assessed to make the model operationally viable. This is consistent with good decision making. If too few options are examined, the decision maker risks omitting a potentially worthwhile option or group of options from the study. In most situations, experience has shown that in or around 10 is the most suitable number of options to include within an evaluation. However, a risk also exists in having too many options, as the decision procedure may become time consuming, overloaded and lacking focus. If too many options are identified at the start of the process, one should check whether some of the chosen options are so closely linked that they are variants of each other, and can therefore in effect be represented by one option for the purposes of decision making. If one refers back to the case study in Chapter 7, it can be seen that in two instances, with B2/B3 and B5/B7, one option was judged to be so close to the other that they were amalgamated for the purposes of the study. Conversely, however, if too few options exist, it may be that not all actors have been consulted. To avoid this, therefore, one should strive to ensure that all actors relevant to the decision process have been consulted.
- The ELECTRE model also ensures that the opinions of the decision-makers are explicitly taken into account within the decision process. This occurs via their agreement both on the values placed within the performance matrix, where each option's score on each of the criteria is given, and on the importance weightings to be assigned to the individual criteria.
- In many instances, ELECTRE forces the decision-maker to select all the characteristics that allow the comparative assessment of options relevant to the decision problem. Only those characteristics whose performance can be assessed on some measurable scale, be it cardinal, ratio or ordinal, can qualify as criteria or sub-criteria of assessment. If it is acceptable to the decision-makers that a set of sub-criteria may, because of their similarity, be combined using a total aggregation method, then this set can form a single criterion. This was the case in Chapter 6 with criterion C3.2, Adaptability to Possible Increases in Waste Production, where each option's performance is estimated on the basis of an index value derived from the weighted aggregation of its score on five separate sub-criteria. In situations where this additive procedure is not possible, they must remain separated, and be given the status of a criterion within the decision problem.

8.2 The Versatility of ELECTRE at Different Planning Levels

The planning of an infrastructure project can take place at many different levels of detail, depending on how far the process itself has been advanced.

If the planning is at a very preliminary stage, with many diverse options still under consideration and with decision-makers making broad strategic choices, the criteria being used to compare the options may be very qualitative in nature. At this level of assessment, there would be relatively little input from engineering and scientific experts on the hard quantitative assessment of economic, technical and environmental criteria.

At a further, more detailed stage in the project planning process, fewer options may be under consideration, with their performance on each of the criteria being considered in more technical detail by the relevant expert groups. At this level of assessment, many of the criterion scores derived may be to some degree dependant on an agreed subjective opinion put forward by the relevant group of specialists, rather than being the result solely of an objective measurement of the performance of the options on the criterion in question. Maystre and Bollinger (1999) term this type of criterion assessment based to some degree on an agreed expert opinion as 'interpersonal subjectivity'.

Finally, an advanced stage in the process can be reached where a definitive, final list of feasible options have been arrived at, and scientific expertise is used, where possible, to achieve objectively based evaluations of the physical effects on each option of the most important criteria.

The case studies in the book illustrate these levels of assessment. In the Mutton Island project in Chapter 5, the assessment carried out was preliminary in nature, and involved the evaluation of some broadly based locational options. It contained no hard quantitative assessment procedures. All criteria were qualitatively based. In the case of the Dublin Port Motorway in Chapter 7, while many of the environmental criteria were qualitatively assessed, their valuations were in each case the result of detailed discussions between the consulting engineers and the relevant experts. In contrast, with the East Switzerland case study in Chapter 6, many of the criteria such as 'waste transportation' and impact of 'gas emissions' are objectively based on quantitative measurements, with only three of the criteria qualitatively based, with the scores agreed by the negotiating group.

The contrast in the level of detail between the first and last case studies mentioned above is echoed in the version of the ELECTRE model used in the respective cases. In the case of Mutton Island, where assessments are preliminary and qualitatively based, one of the simplest version of the model – ELECTRE II – is used. For the East Switzerland Study, with detailed criterion information available, the more complex ELECTRE III model is employed. It should be noted that, for a given case study, different versions of the model can be used as one moves through the planning process and the available information becomes more detailed and focused. One could also revert back to a version of the model used previously at an earlier stage in the process. Within the decision problem, this flexibility of movement between versions contributes to a robustness in the final result.

When planning an infrastructure project, the three case studies highlight the centrality of environmental criteria to the option choice process. The OECD (1994) put forward a framework for presenting information on environmental criteria. They detailed three levels on which this information could be provided:

Level 3
A Parameter – A property of the criterion in question that is measured and observed

Level 2
An Index – An set of aggregated or weighted parameters

Level 1
An Indicator – A value that provides information about, and describes the state of, a phenomenon. It is general in nature, extending beyond that which is directly associated with any one parameter value.

The level at which the criterion information is given will determine the version of ELECTRE to be used. The simpler versions of ELECTRE – ELECTRE I and II, require information to be provided at the most basic Level 1. Data at this level is usually predominantly qualitative in nature. The Mutton Island case study in Chapter 5 is an example of this. All criteria are general in nature, expressing on a set graded scale each option's overall performance in areas such as visual impact, effect on birdlife and construction impact. The criterion valuations are based on the interpersonal subjectivity of the decision group members. As the information conforms more to Levels 2 and 3, the more complex forms of the model – ELECTRE III and IV – become more appropriate. Within the East Switzerland case study in Chapter 6, all the environmental criteria are either individual parameters such as Energy Use which can be quantified in GigaWatt hours per year for each strategic option, or weighted parameters such as Adaptability to Possible Increases in Waste Production which is, as noted above, derived from the weighted aggregation of five parameters. Two other criteria - Adaptability to Possible Decreases in Waste Production and Overcapacity are also estimated using such indices of aggregated parameters. In contrast, The Dublin Port Motorway case study, the two quantitatively assessed environmental criteria – noise and air pollution - use the measurement of a single parameter to signify the extent of the criterion impact on a given option. The Air pollution effect is expressed solely in terms of Carbon Monoxide concentrations in parts per million, with other less important parameters such as particulates and oxides of Nitrogen are not considered. Noise pollution expressed in decibels in terms of increases in traffic noise. The additional effect of vibration due to traffic is not considered. The use of one parameter only is justified by the preliminary nature of the study. Thus, one of the great strengths of ELECTRE lies in its ability to adapt to its use at different stages in the project planning process, and its ability to cope with information provided in different forms and in different levels of detail.

8.3 ELECTRE and the Decision Makers

The outcome or results of any decision process will be worthless if any actor, or group of actors, rejects the recommendations put forward, on the basis that their views have not been reflected within the process. Their exclusion, perceived or otherwise, will have an adverse effect on the credibility of the procedure. Therefore, the composition of the team of actors that comprise the negotiating group working with the decision model will have a direct bearing on the level of acceptance the final result will receive. The group must encompass all relevant legitimate opinions. Each actor uses criteria of evaluation to express, in a rational manner, his or her opinion regarding the relative merit of each of the options under consideration. Because these criteria cannot be reduced to a single unique one, the problem becomes multi-criteria based.

The ELECTRE Model, as stated initially in Chapter 1, lies at the heart of the 'compromise principle'. Within a multi-criteria framework, it provides a 'vehicle' for expressing the points of view of all the participants within the decision process. It thus helps record, in a simple and effective way, the results of the discussions, disagreements and compromises between the major actors. Furthermore, the ELECTRE model allows the conversion of a succession of 'bargainings' into a group learning process, in an explicit and open way, contributing to the coherence of the final result.

The three case studies in the previous chapters have illustrated that areas of incomparability within the data are shown explicitly in the final pre-order by the parallel ranking of such options. This contributes to the coherent interpretation of results. A relative ranking is not forced on any two options where insufficient information exists to make a direct connection between them.

Within the East Switzerland case study, where more than one decision-maker exists, the ranking of all options from the perspective of each of the individual decision makers is used as a basis for the final decision. Average importance weightings are only employed as a basis for carrying out sensitivity and robustness testing on the baseline results. In terms of the final decision, however, no attempt is made to submerge the individual rankings into an overall agreed ranking, which may well not be the actual opinion of any of the actors involved. The final decision thus attempts to order the options on the basis of some degree of consensus or disagreement between the rankings of the individual decision-makers. In the Eastern Switzerland case, four options were shortlisted on the basis of their consistently high ranking among all actors, with one option featuring particularly prominently.

8.4 The SURMESURE Methodology and Decision Makers

The SURMESURE Technique, put forward by Pictet, Maystre and Simos (1994), is a two-dimensional diagram that illustrates the upward and downward distillation rankings of an option obtained from the ELECTRE model. For a given decision-maker, as shown in Figure 8.1, a single point represents the coincidence of these two rankings. In those situations where more than one decision-maker is directly involved in the process, SURMESURE allows the results of ELECTRE for a given option, from the perspective of each of the decision-makers, to be illustrated in an easily understandable fashion. It can thus be of great use in estimating the degree of consensus among the decision-makers regarding the ranking of a given option relative to the others under consideration.

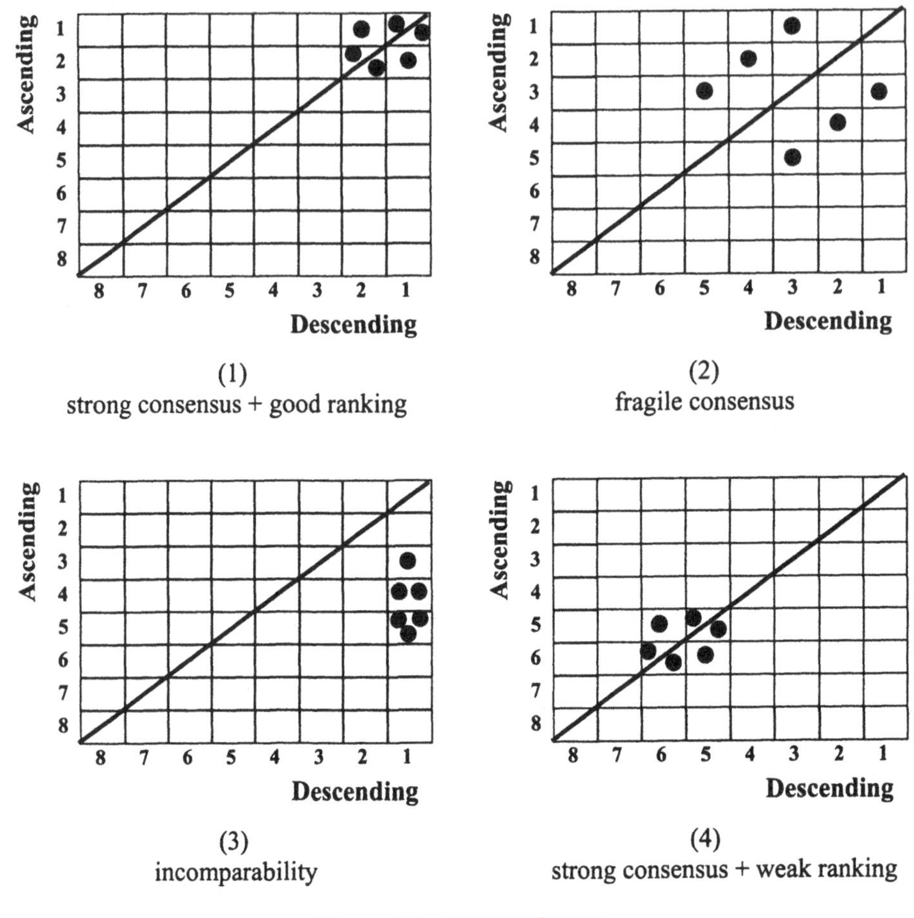

Figure 8-1 – SURMESURE

For each of the options under examination, the ascending distillation score is given on the vertical axis, with the descending distillation score on the horizontal axis. In the above diagram, we have assumed six decision-makers. The four diagrams above indicate the main features that the SURMESURE Diagram illustrates:

a) The strength or fragility of the consensus regarding the options ranking,

b) The level of incomparability in the option relative to the others under consideration by the decision-makers, and

c) The actual ranking of the option.

These features are illustrated using the four diagrams in Figure 8-1 as follows:

1) The consensus regarding the ranking of the option in question is illustrated by the close clustering of the points. The proximity of all points to the diagonal indicates the high level of comparability between it and the other options. The closeness of the cluster of points to the top right hand corner of the diagram indicates a high ranking (ideally, all decision makers would rank the best option first in both ascending and descending distillations).
2) The dispersed nature of the points indicates a lack of consensus among the decision-makers regard the ranking of the option. All points are somewhat distant from the diagonal, indicating a degree of incomparability between it and the other options. While some of the points are close to the top right hand corner, others are more distant, indicating a variation in ranking among the decision-makers from good to moderate.
3) In this case, the distance of the points from the diagonal indicates a high level of incomparability between it and the other options. The close clustering of the points indicates a high level of consensus regarding the absence of comparisons.
4) This result is similar to 1) in that the close clustering of the points and their proximity to the diagonal indicates high levels of consensus and comparability respectively. Unlike the first case, however, the distance from the top right hand corner indicates an at best moderate ranking for the option in question.

In overall terms, the technique provides a basis for deciding on the general ranking of each option relative to all others, the degree of consensus and clarity among the decision-makers regarding the ranking.

8.5 The Assessment of Qualitative Criteria

Introduction

While it is always preferable to assess the decision criteria in some quantitative manner, a proportion of them will always require a qualitative appraisal. All case studies referred to in previous chapters contain criteria that are qualitatively based. Given their existence, it is necessary to put forward a coherent and understandable methodology for their estimation.

Basic Qualitative Comparison of Project Options

At the most basic level of assessment, for a given criterion, a group of options can be compared on a nominal scale by assessing the level of similarity / dissimilarity between them. It is a very fundamental and indeed universal desire for humans to compare and contrast such objects to see how alike or different they are, and, as a consequence, to have some sense of their similarity or dissimilarity. This type of analysis is independent of preference relationships. One could more readily describe this procedure as nominal scaling, which is the first and simplest level of analysis, preceding the use of interval, ratio or cardinal scales.

This first level of analysis is a useful tool for building what might be termed categories, families, alliances or clusters of options on a given criterion. It provides a simple basic picture of the relative positions of the options being considered within the context of the objective of the decision process as expressed by the criterion under consideration, be it the minimising of landscape effects or the maximising of political acceptability. This nominal evaluation is a useful preliminary tool of comparative assessment.

Moving to the next level of analysis, one can describe the level of similarity between two options on a given criterion on an ordinal scale. One proposed by Maystre (1999) has five grades, comprising the extremes, the median grade, and that which lies between the middle and both extremes. These five grades can be expressed semantically as follows:

- Very similar
- Fairly similar
- Moderately similar
- Slightly similar
- Not similar

As a graphical illustration of this scale, Table 8-1 indicates the pairwise levels of similarity between nine options on the above scale. For those that are considered very similar, the cell of the matrix is coloured black, with the cells shaded progressively brighter as the relationships become less similar. The cells for those pairs judged not similar are coloured white. Table 8-2 gathers all like-graded options together.

	A1	A2	A3	A4	A5	A6	A7	A8	A9
A1									
A2									
A3									
A4									
A5									
A6									
A7									
A8									
A9									

Table 8-1

	A1	A8	A7	A5	A2	A4	A9	A3	A6
A1									
A8									
A7									
A5									
A2									
A4									
A9									
A3									
A6									

Table 8-2

This type of preliminary information will help enrich the exchanges between decision-makers. It forms an important basis for the assessment of qualitative criteria within the ELECTRE model. Two such methods are:

- The Methodology of Rogers and Bruen (1998)
- MACBETH (Bana e Costa and Vansnick, (1997))

8.6 Assessing Qualitative Criteria within ELECTRE (Rogers and Bruen, 1998)

Within the context of option choice problems in the planning of major infrastructure projects, the performance of each of the proposed projects relative to the existing situation or 'status quo' is of vital importance, since it confirms the worth of proceeding with the project in question. As with the Maystre technique referred to above, a five-point scale is utilised. However, within this technique, for a given qualitative criterion, the scale indicates level of similarity between each option's performance and that of the status quo as estimated by the decision-makers. This five-point scale for the evaluation of each option relative to the existing situation is as follows:

- Neutral / indistinguishable
- Minor (adverse or beneficial)
- Moderate (adverse or beneficial)
- Major (adverse or beneficial)
- Severe (adverse only)

Because almost all major versions of ELECTRE involve the pairwise comparison of options, each option's score on the above scale is used as a basis for deciding whether the relationship between the two is one of indifference, weak preference, strict preference or veto inducing. The results from it thus become completely compatible with those pairwise outranking relationships derived for each of the quantitatively based criteria. The operation of this ordinal scale is detailed within the ELECTRE IV case study in Chapter 7.

This ordinal scale is assumed to be linear, i.e. the distance between successive points is equal.

Rogers and Bruen (1998) also put forward a second system where, for the qualitatively based criterion in question, the options were compared on a direct pairwise basis rather than on the basis of their respective performances relative to the existing situation. This strictly pairwise system was devised for a number of reasons. Firstly, there may be situations where the 'do-nothing' option does not exist or cannot be defined. In addition, the pairwise system has been found to have certain advantages. Saaty (1980) believed that, in situations where a decision-maker had a significant level of expertise, there was 'no better way' of getting those judgements down than through a systematic procedure, such as that provided by such a system. Furthermore, Saaty found the pairwise comparison approach to have a level of consistency not available to techniques where each option is compared to a baseline case rather than directly to each other. The information gathering process is more arduous. The information obtained from the procedure will, however, be more complete, allowing more definitive statements to be made regarding those pairs of options whose preference relationship may not be clear from the results of the first method. The six point scale used for the direct pairwise technique is as follows:

1) there is **no difference** between the two options
2) there is a **just noticeable difference** between the two options
3) there is a **noticeable** but not definitive difference between the two options, (indicating weak preference for one over the other)
4) a **clear**, strong preference can be expressed in favour of one option over the other
5) a **very strong** / substantial preference can be expressed in favour of one option over the other
6) there is an overwhelming difference between the two options, indicating **absolute** preference for one over the other (indicating a veto against the worse performing option)

This scale allows indifference, weak preference strict preference and veto enforcement to be directly inferred from the relative performance of any two options on the above scale.

The two methods can be used together. If the study is relatively preliminary in nature, as was the situation with the case study in Chapter 7, the first method can be used to get a broad picture of the more obvious preference relationships. The second method could subsequently be used on a more focused group of options whose preference relationships could not be ascertained with confidence from the first technique.

8.6 MACBETH (Bana e Costa and Vansnick, 1997)

The choice of evaluation scales for criteria that are either cardinal or ordinal in nature is relatively straightforward. In the case of a cardinal scale, the option is given a numerical value on the appropriate scale, while ordinal criteria are constructed in a straightforward manner by assigning the option to one of a number of appropriate gradings. The two-stage technique of Rogers and Bruen (1998) outlined above uses an ordinal scale to assess all options on the criterion in question.

However, certain criteria sometimes require ordinal scales of evaluation that progress in a non-linear fashion. The scale in question is neither purely cardinal, involving a numerical evaluation, or purely ordinal, involving a simple classification. This type of hybrid scale is used to establish an order of preference that includes a notion of distance between the different ordinal classifications. The method of construction of such scales is complex. One must define the distances between different preference classes. While certain experts may be capable of defining these distances between points on such a scale, in most situations this would not be feasible.

MACBETH is a methodology that permits the pairwise qualitative assessment of options. It allows a decision-maker or group of decision-makers to be guided in the construction of a numerically based interval scale within which the notion of difference in degree of preference is significant. It thus obtains preference information richer than simple ordinal judgements.

The development of the method is based on the assumption that an actor is capable of formulating with words as well as with numbers the strength with which he prefers one option relative to another. MACBETH thus permits comparison in a global / overall way, of one option with another. However, when using it in order to obtain a rank ordering of options based on the relative preference relationships between them, all the options must be comparable in the eyes of the actor being questioned. For decision problems within the areas of engineering and infrastructure investment, particularly where environmental concerns are significant, full comparability between options is seldom achievable. Hence, in the context of this book, this method is not at significant for the global comparison of different scenarios and options as the ELECTRE Methods detailed in previous chapters.

Nevertheless, on any given criterion, the MACBETH approach has been found very useful for building a non-linear ordinal scale of evaluation, allocating numerical values to different reference levels. It can be termed an interactive approach, developed on the basis of the following four fundamental points:

1) MACBETH helps an actor, or group of actors, produce information richer in nature than the simple ordinal form. It adopts a mode of questioning very simple, comparing only one pair of options at any time, and requiring solely a verbal response. It becomes necessary, therefore, to pose numerous questions in order to be able to test the compatibility of the responses with the construction of a cardinal scale.

2) In the cases where such incompatibilities arise, one must be capable of detecting their possible source in such a way as to be able to initiate a serious discussion

with the relevant actors and help revise their initial judgements to obtain full compatibility between responses.

3) In cases where all responses are compatible, it is important to propose an initial numerical scale on the basis of simple principles and rules (rules of measurement consistent with the semantically based information provided by the actors concerned).
4) A user friendly software is required which both allows an easy understanding of the MACBETH scale and expresses significance in terms of the difference in attractiveness between the options, transforming these differences onto a cardinal scale

The basic procedure for MACBETH is stated as follows

- For a given criterion, the actors within a negotiating group allocate each pair of options to one of a number of reference levels, each of which is defined in order of decreasing attractiveness;
- In order to 'fine-tune' this classification these actors seek to attribute a score to each one of these reference levels in such a way that the scale derived is cardinal.

To help the actors move beyond the ordinal scale, the framework of MACBETH requires the decision-makers to allocate the difference in attractiveness between any given pair of options into one of the following six categories:

1) Very weak difference in attractiveness (very weak preference in favour of one option)
2) Weak difference in attractiveness (weak preference in favour of one option)
3) Moderate difference in attractiveness (moderate preference for the preferred option)
4) Strong difference in attractiveness (strong preference for the preferred option)
5) Very strong difference in attractiveness (very strong preference for the preferred option)
6) Extreme difference in attractiveness (extreme preference for the preferred option)

The following type can be used in order to review this type of information:

The number of semantic expressions – seven - is set within MACBETH. It falls within the range five to nine, the number of grades generally seen as sufficient to express the level of preference or attractiveness judged to exist between two objects. Within the context of a decision problem, therefore, an actor must assign the preference relationship between each pair of options, for a given criterion, into one of above the six categories or must decide on a zero difference in attractiveness. One can then construct a matrix of judgements.

As soon as the table of judgements is introduced into MACBETH, the software tests whether the information given is compatible with the construction of a cardinal scale. In the case of Table 8-3, the information does permit the construction of such a scale. It is thus possible, through the refinement of the data, but without altering it, to arrive at cardinally based output. MACBETH proposes a coarse numerical scale which is consistent with the data on the options provided by the group of actors. Figure 8.2 below represents output from the MACBETH software for the Table of preference judgements in Table 8.3

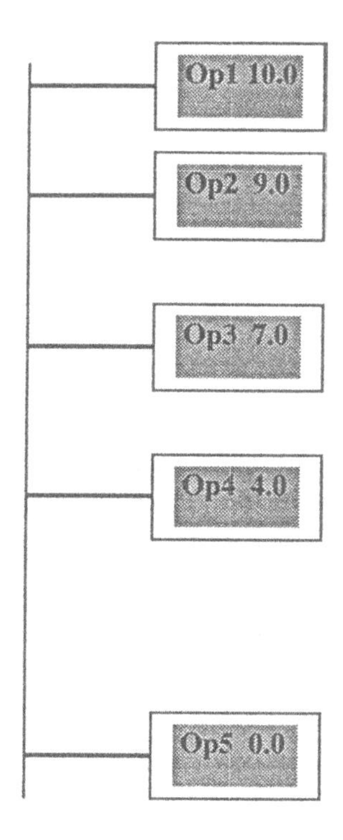

Figure 8-2

	Option 1	Option 2	Option 3	Option 4	Option 5
Option 1	None	Very weak	Moderate	Strong	Extreme
Option 2		None	Weak	Strong	Very Strong
Option 3			None	Moderate	Strong
Option 4				None	Strong
Option 5					None

Table 8-3 – Performance Table for 5 Sample Options

Experience has shown that if the decision-makers are faced with more than five options, they begin to have difficulties in expressing consistent relationships between each of the pairs to be considered. In such a situation, it is simpler and faster for them to complete, in the first instance, those comparisons on the diagonal of the matrix as shown in Table 8-4, where each neighbouring pair of options is successively compared. The other comparisons can then be completed in a simple and logical manner.

	Option 1	Option 2	Option 3	Option 4	Option 5
Option 1					
Option 2					
Option 3					
Option 4					
Option 5					

Table 8-4

The more the options to be compared are different in nature, the more their systematic comparison becomes difficult and provokes resistance on the part of the actors.

8.7 AGATHA

When one compares a full set of project options using a decision model, a final ranking or total pre-order is obtained. It is important to consider, however, whether the same ranking would have been obtained if one or indeed a number of the original options had been omitted from the evaluation and only a sub-group of the original set had been considered. One can express this more precisely using the following example. A decision-maker produces a ranking for a set of twenty-two options using the appropriate ELECTRE model. If the decision-maker then omits the worst ten performing options, would the four or five best performing options under the original ranking from the full list of options be in identical positions within the new ranking for the reduced set of twelve? Maystre and Bollinger (1999) have devised the following procedure, called AGATHA, for checking the robustness of the final ranking as the worst performing options are progressively eliminated. It is detailed as follows:

- The ranking of options by ELECTRE takes place at every step. For each of the options, the sum of its upward and downward distillations is calculated. If more than one actor is involved, the score for each of the actors is summed to give a final tally for each option. The numbers are then normalised, with the worst performing option assigned the score 1, corresponding the poorest rank.
- On a two-axis graph, the number of options being compared is placed on the horizontal axis, with the normalised sum of the ranks on the vertical axis.
- Having eliminated the worst performing option, i.e. the one having the normalised score of one, the ranking procedure is repeated, and the normalised ranking scores are calculated for the remaining options.
- A graphic is obtained on which every option is represented by a broken line that starts at the point on the graph where all options are being considered and ends when it is assigned the normalised ranking score of 1.
- If there is no location at which the broken lines of any two options intersect, then the decision-maker can confirm that the ranking obtained by the model is independent of the number of options considered.

Figure 8-3 below illustrates this procedure with 5 options. One can see that Option 5 has the poorest ranking on the initial run of the model when all five are being assessed. It thus appears as a point. The ranking of Options 1 and 4 are independent of the total number of options considered. The broken lines corresponding to these two options never cross over each other, and Option 1 is always ranked the lower of the two. In contrast, Options 2 and 3 are dependent on Option 4. Their relative ranking is reversed once Option 4 is removed More detailed evaluation and sensitivity testing would have to be carried out on Option 2 to confirm or contradict the second place ranking of Option 2 obtained from the initial run of the model with the full five options. Its position in the different ranking is manifestly dependent on the presence of Option 4. In contrast, the performance of Option 1 relative to the others is such that should clearly be ranked first.

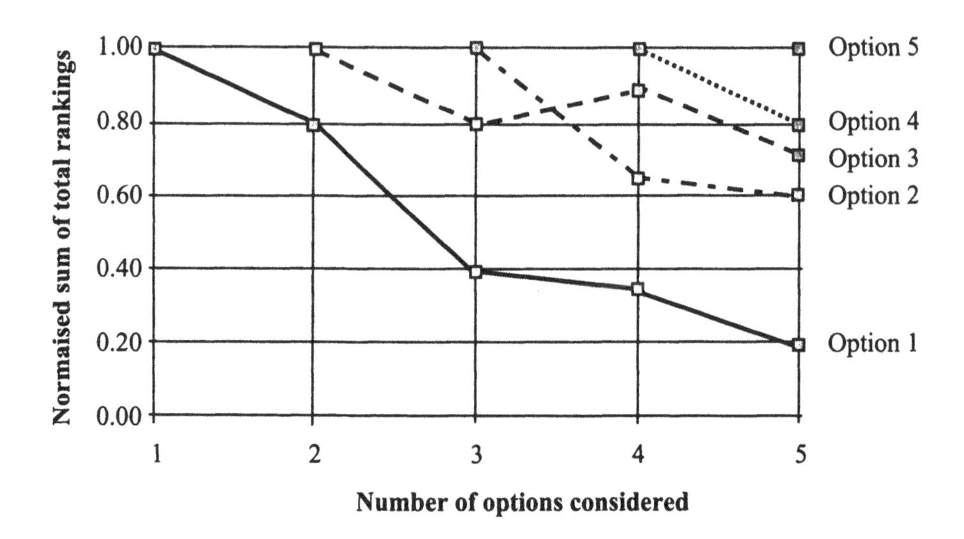

Figure 8-3 – AGATHA : A graphical illustration of the progressive reduction of comparisons

AGATHA shares many of the advantages of the SURMESURE procedure. It is graphical, global and dynamic all at once. It is, however, merely a graphical representation of the relative positions of the options. The physical distance separating the options on Figure 8-3 is not an indication of their relative performance. It is an ordinal scale, not a ratio one.

8.8 Available Software

ELECTRE software is available from The LAMSADE Laboratory at the University of Paris, Dauphine. At the LAMSADE website, trial versions of all the major models are available for downloading.

The address of the LAMSADE website is as follows:

http://lamsade.lamsade.dauphine.fr/english/software.html

Full versions of the software are available for purchase from LAMSADE.
A trial version of the MACBETH software can be downloaded from the MACBETH website. Full versions of the MACBETH software can be purchased from either of the authors.

Software for the revised Simos weighting technique (SRF) referred to in Chapter 4 is available from Jose Figueira, University of Coimbra, Portugal.

8.9 References

Bana e Costa, C. and Vansnick, J.C. (1997) 'Applications of the MACBETH Approach in the Framework of an Additive Aggregation Model' *Journal of Multicriteria Decision Analysis*, 6:2, pp107-114.

Maystre, L.Y. and Bollinger, D. (1999) *Aide a la Negotiation Multicritere: Pratique et Conseils*. Presses Polytechniques et Universitaires Romandes.

OECD (1994) *Environmental Indicators / Indicateurs de l'Environnement*. Organisation for Economic Co-operation and Development.

Pictet, J, Maystre, L.Y. and Simos, J. (1994) "SURMESURE: An Instrument for Representation and Interpretation of ELECTRE and PROMETHEE Method Results", in Paruccini (ed), *Applying Multiple Criteria Aid for Decision toEnvironmental Management*, Kluwer Academic Publishers, Collection "Eurocourses", Dordrecht, pp. 291-304.

Rogers, M.G. and Bruen, M.P. (1995). "Non-Monetary Based Decision-Aid Techniques in EIA – An Overview". *Proceedings of the Institution of Civil Engineers, Municipal Engineer*, Vol. 109, pp98-103, June.

Rogers M.G., and Bruen, M.P. (1998) "Qualitatively Assessed Criteria within Outranking methods". Stream Title - Qualitative assessment within MCDA, Session MD, *EURO XVI, 16th European Conference on Operational Research, Brussels*, 12th to 15th July.

Saaty, T. (1980) *The Analytic Hierarchy Process*. McGraw Hill.

Index